The City Project

Strategies for Smart and Wise Sustainable Urban Design

Volume 6

The book series The City Project reports on applied research and operational developments that promote urban renewal and the sustainable transformation of contemporary cities. Inspired by the "City of Man" as imagined by Adriano Olivetti and Ernesto Nathan Rogers, and going beyond the concept of the smart city and related technological advances, the series' goal is to present holistic, practice-oriented and multidisciplinary strategies for realizing the City 4.0, i.e., the city of the fourth industrial revolution, in keeping with the objectives of the UN's 2030 Agenda for Sustainable Development.

In particular, the series reports on effective design, planning and management approaches that leverage urban and architectural design skills, engineering, environmental and social expertise, and administrative abilities alike. It welcomes books on each of the aspects mentioned above, as well as studies analyzing multiple aspects, their interactions and/or holistic solutions. The City Project addresses a very broad readership, including designers, engineers, architects, social scientists, stakeholders and public administrators, who deal with various aspects of the realization of the City 4.0. It publishes theoretical investigations into the contemporary built environment, international case studies, and pilot projects concerning urban renewal and the regeneration of urban areas, as well as the proceedings of key international conferences.

Books published in this series are devoted to supporting education, professional training and public administration. Outstanding PhD theses on emerging topics, if properly reworked, may also be considered for publication. The series is published with the support of the Smart City 4.0 Sustainable LAB, an interdisciplinary teaching and research project on future cities initiated by the University of Parma, and jointly implemented with other regional universities (the University of Bologna, University of Ferrara, and University of Modena and Reggio Emilia).

About the Cover

The cover of the book series The City Project features a painting by Carlo Mattioli (C. Mattioli, *Estate in Versilia*, 1974, oil on canvas cm. 118 × 70, Catalog n. 1974D0029, Courtesy of Fondazione Carlo Mattioli, thanks to Anna Zaniboni Mattioli)

The horizon of poppies painted by Carlo Mattioli between the dark background of the forest and the white plane of the wheat, becomes for us, thanks to a transfiguration of meaning that aligns with the attitude towards abstraction rooted in the figure of the painter, a city which is intertwined with its landscape, evoking the idea and the possibility of recomposing a balance and seeking an integration between settlement and environment, between human space and natural element.

Dario Costi · Giovanni Leoni
Editors

Smart City: A Critical Assessment

Springer

Editors
Dario Costi
Department of Engineering
and Architecture
University of Parma
Parma, Italy

Giovanni Leoni
Department of Architecture
University of Bologna
Bologna, Italy

ISSN 2730-6992 ISSN 2730-700X (electronic)
The City Project
Strategies for Smart and Wise Sustainable Urban Design
ISBN 978-3-031-51287-2 ISBN 978-3-031-51288-9 (eBook)
https://doi.org/10.1007/978-3-031-51288-9

Cover illustration: C. MATTIOLI, Estate in Versilia, 1974, oil on canvas cm. 118 × 70, Catalog n. 1974D0029, Courtesy of Fondazione Carlo Mattioli, thanks to Anna Zaniboni Mattioli.

This Springer imprint is published by the registered company Springer Nature Switzerland AG
The registered company address is: Gewerbestrasse 11, 6330 Cham, Switzerland

Paper in this product is recyclable.

Foreword

The Smart City Idea of the Emilia-Romagna Region

The cultural debate on the theme of the "smart" city is an opportunity to reflect on this adjective, which is widely accepted uncritically—as we are now used to doing for other "slogans"—without considering its real, actual meaning. The role of cities in the development processes of a territory is a topic widely debated by the Emilia-Romagna Region in defining its Regional Energy Plan 2030. The city is a topic of absolute priority for those who deal with energy and climate, since 70% of the European population lives in cities and, for this reason, urban areas contribute significantly to energy consumption and greenhouse gas emissions. At the same time, cities represent the main engines of the European economy and in the Pact for Work and Climate—which the Emilia-Romagna Region signed in December 2020 with more than 55 subjects, associations, companies, institutions, universities, professional orders and colleges—the city's role has been recognised in contributing to the Region's regeneration and development project based on sustainability through its inseparable components: environmental, social and economic.

The definition of a city's role within an ecological transition pathway capable of reaching the goal of climate neutrality is a broad and interdisciplinary topic. Climate neutrality means dealing with urban planning, energy, mobility and transport plans, controlling emissions, increasing the energy efficiency of buildings and installations, managing data, networks and services and social relations in order to manage a just and inclusive transition. This complexity has been exacerbated by the COVID-19 emergency, which brought about changes that have made the need to make the city "smarter" even more evident and urgent. In this context, smarter means a better and fairer spatial and social entity, capable of rekindling communities' interest in greater "resilience", understood as the capacity to adapt both to health emergencies and to those induced by climate change or other events capable of bringing about substantial changes to the quality of life.

To address this complex issue, in 2019, the Emilia-Romagna Region signed a protocol with the University of Parma and LEPIDA to implement a project aimed at expanding the *Smart City 4.0 Sustainable LAB* research laboratory to develop a research and training centre on the themes of smart and sustainable cities with the aim of extending the boundaries of the "Smart

City" to the entire regional territory and contributing to demographic, productive and economic rebalancing. In particular, we believe that the laboratory can serve as a research space capable of providing guidelines for urban regeneration practices and, more specifically, for network upgrading processes, for rationalising the use of resources, for identifying measures capable of tackling climate change effectively, for promoting eco-innovation and the "green economy", and for defining policies to reduce social and economic inequality. For this reason, the laboratory has brought experts from different fields together around each of the various projects, including environmental scholars, biologists, economists, engineers, architects and anthropologists. The key focus for the laboratory has been "participation", because everyone can make a contribution, because the city belongs to everyone.

The following publication provides a Glossary for the smart, wise or, even better, the just city, which will provide an opportunity to reflect on the future of places where the integration between knowledge, technologically advanced structures and services can contribute to sustainable growth, wellbeing and improved quality of life; a city capable of producing new forms of social cohesion and linking material infrastructure with human and social capital.

If we are talking about New Technologies, we must be able to ensure that everyone can make use of them. Otherwise, instead of being a positive element, they too could become a cause of social exclusion and marginalisation. We will have to be able to educate citizens in terms of awareness, knowledge, competence, interpersonal skills, inclusive attitudes, improving established behaviours and relationships and implementing participatory planning methods that allow everyone to perceive real democracy in relation to the decisions that affect them.

Cities of the future will be conceived and designed as hyper-connected ecosystems, dotted with sensors and devices capable of collecting and processing an enormous amount of data that can, if handled wisely, contribute to the management of urban functions (energy, security, mobility, health, residence, study, work, production, etc.).

Another topic of great interest is the expansion of the Smart City to the entire regional territory, an operation that requires deep reflection on the relationship between urban life and life in the suburbs or provinces. The distribution of services and functions will have to be rebalanced through targeted investments aimed at reversing the process of impoverishment that has affected many peripheral areas and led to the loss of human capital in those places.

A final issue I would like to highlight is that of training, an essential activity for local authorities that will be required to have skills and abilities that they do not possess today. These professional skills can contribute towards strengthening active citizenship, increasing the capacity to build strong communities and facilitating the identification of strategic locations for urban development capable of attracting market players interested in investing in the implementation of interventions.

In parallel with the laboratory's work, the Region has launched a series of actions to help make its territories and urban areas more attractive and sustainable places to live in by immediately applying the concepts of Smart Cities:

- promoting cleaner and more sustainable public and private mobility through support for the implementation of Sustainable Urban Mobility Plans (SUMPs), the renewal of Local Public Transport (LPT) fleets and regional trains, the introduction of systems to increase the attractiveness of LPT (infomobility, mobile ticketing, video surveillance on board vehicles and at stops), intermodality, the improvement of goods logistics, the creation of cycle paths and the spread of charging stations for electric mobility;
- encouraging the energy requalification of public buildings;
- simplifying procedures for the renovation of private buildings;
- supporting Sustainable Energy and Climate Action Plans (SECAPs);
- trialling energy communities.

These are merely the first steps of a long journey we have undertaken to contribute to change in our cities, fully aware that we will certainly have to take other actions since there are many variables to take into account. The authoritative testimonies presented in the publication will help us understand how to make the concept of the sustainable city a reality, focusing on all those specific concerns that will enable us to move in the right direction

Attilio Raimondi
Responsible for the implementation of the Regional
Energy Plan, Research, Innovation,
Energy and Sustainable, Economy Service
Emilia-Romagna Region
Bologna, Italy
Attilio.Raimondi@regione.emilia-romagna.it

Preface

The following volume is the first testimony of a multidisciplinary collective work focused on the topic of the Smart City from both scientific and operational points of view.

The intention is not to passively assess the topic but rather subject it to criticism in order to relaunch its current and productive aspects, also updating its terminology. The reflection on the terminology, which in the text takes the form of a reasoned glossary, does not derive from requirements of pure philology but rather aims to make the discussion among the disciplines more grounded and specific in order to build a field of research-action with a wide spectrum of skills involving the various actors—specialists and otherwise—that have an interest in the topic. On the one hand, an effort to reunite the current narrative of the Smart City and its variations (Wise City, Wise Town, Care City, 15 Minute City, etc.), and on the other the real city, no less elusive in its persistence and transformation with powerful accelerations such as the one triggered by the pandemic.

Several aspects of the study's design and launch delineate the orientations of this update.

1. The geographical reference to the Emilia-Romagna Region, to its cities that recall the specific historical city of European origin: a place of elaboration of the conflict of cohabitation; an example of fruitful intermingling of private and public spaces: a laboratory of community and mobility that becomes social. A model that sheds light on the false nature of the alleged conflict between tradition and innovation and offers numerous opportunities to combine cultural identities and technological experiments to their mutual advantage.
2. The assumption of the Smart City not as an abstract concept or model but as a specific process characterised by conscious convergence and interdependence between humanistic reflection, technological innovation, assessment of economic and social impacts, policy orientations. A continuous process, as is continuous the transformation of cities, substantiated by education and research, able to reconnect the fragmented multiplicity of city plans and to overcome political vagaries pushing in the direction of the simplification and fluidification that are required by the rapid pace of innovation.

3. The choice of urban culture as the target of reference in all its various facets and compared with the fourth industrial revolution. Consequently, the assumption of the Smart City as a project that is as complex as its area of application: the city. Thus is defined a terrain for the comparison of different types of knowledge that are seen as inter-connected and inter-enabling, without predetermined hierarchies: cultural elaboration of the city and its images; cultures of the project of transformation, conservation and development of the city; energy issues; technological innovations; artistic practices and creative industries.
4. The desire to reconnect this process to the physical existence of specific, real places inhabited by specific, real communities by extending this attention to the physicality of both the visible city of the surface and the invisible city of the technological networks underground or in the air, having as its objective the search for a joint, reciprocally measured functioning of the two.
5. The objective of subordinating technological applications and purely economic aspects to criteria of benefit—in the short as well as in the long term—for the communities of reference by introducing principles of social economy and social inclusion to the field of Smart City research-action.

Parma, Italy — Dario Costi
Bologna, Italy — Giovanni Leoni

About the Laboratory Research Team

Smart City 4.0 Sustainable LAB is a laboratory carrying out research on the contemporary city at the University of Parma. It has been supporting public bodies, institutions and stakeholders in the territory since 26 March 2018. In 2019, on the strength of a memorandum of understanding signed between the University of Parma, the Emilia-Romagna Region and Lepida ScpA, the laboratory formed an interdisciplinary and inter-university network composed of professors from the Universities of Parma, Modena and Reggio Emilia, Bologna, Ferrara, and the Milan Catholic and Polytechnic Universities, Piacenza campus. Since then, the research groups of the universities involved have initiated integrated research activity with the perspective of sharing an overall, organic and heterogeneous response that provides solutions for the implementation of intelligent and sustainable cities by collaborating with the regional Federations of Professional Orders of Engineers, Architects, Surveyors and Experts, ANCE (National Association of Building Constructors), ANCI (National Association of Italian Municipalities), CNA Emilia-Romagna (National Confederation of Craftsmen and Small and Medium Enterprises), Legacoop Emilia-Romagna, Arpae (Regional Agency for Prevention, Environment and Energy of Emilia-Romagna), ADBPO (Po River District Basin Authority), Regional Agency for Territorial Safety, Civil Protection and AIPo (Interregional Agency for the Po River).

Laboratorio di Ricerca
SMART CITY 4.0
sustainable LAB

About the Publication

The following publication is the outcome of an invitation-based international conference organised by *Smart City 4.0 Sustainable LAB* on 25 June 2021 entitled *Smart City? A critical position. Glossary and interdisciplinary views. A cultural exchange after a year's work* which had the following structure:

- Presentation and discussion of the research *Glossary of contemporary city terms*, which involved the collection and critical analysis of a series of definitions and neologisms related to some models of organisation and development of the city today.
- Presentation of *Video interviews on the city of the Fourth Industrial Revolution* to Stefano Zamagni, Derrick De Kerckhove, Alberto Broggi, Giovanni Maria Flick, Sébastien Marot edited by the members of the Scientific Committee of the *Smart City 4.0 Sustainable LAB*.
- The critical positioning towards the Smart City topic of the members of the Scientific Committee of the *Smart City 4.0 Sustainable LAB*.

All listed interventions, following the discussion at the conference, were later transformed into research papers and were finally published after positive evaluation by the Editorial Board of *The City Project* Series.

About the Images

The book hosts a photographic research by **Alessandra Chemollo** entitled Along the Via Emilia. Thanks to the choice of an intermediate point of view between the ground and aerial views—in fact photographing from a height of a few metres thanks to the help of the Fire Brigade—and choosing—after the 700 km Modena-Graz exhibition by Franco Vaccari (1972)—a scanning of the images dictated by a recurring measurement in kilometres, Chemollo succeeds in narrating the territories and cities encountered according to a non-selective vision that tells of their complexity, stratification and contradictions. The images, therefore, rather than accompanying the position papers of the members of the Laboratory's Scientific Committee, also express a critical position of their own.

Contents

Editors and Contributors

About the Editors

Dario Costi, Arch. Ph.D. is Full Professor in Architectural and Urban Design in the Department of Engineering and Architecture at the University of Parma. He is Director of the interdisciplinary teaching and research project "Smart City 4.0 Sustainable LAB", which is promoted by the University of Parma in cooperation with the Emilia-Romagna regional government and the public digital infrastructure Lepida and involves the Universities of Ferrara, Bologna, Modena and Reggio-Emila, Piacenza as well as local Authorities and stakeholders. He is a member of the Ph.D. Program in Civil Engineering and Architecture at the University of Parma and at the Faculties of Architecture of the University of Palermo and Sapienza University of Rome. He is active both as a professional and as a researcher in the architectural and urban design field. He is the co-owner of the MC2 Architecture Studio in Parma. Since 2007, he has been serving as President of Parma Urban Center, of which he is also the co-founder. Involved in numerous national projects for fostering education in architecture and urban design, Prof. Costi is the author of many Italian publications and an editor of two Italian book series published by MUP Editore, Parma. Since 2020, he has been the editor of the international book series "The City Project—Strategies for Smart and Wise Sustainable Urban Design", published by Springer. He is also serving as a lecturer for the Program Master Housing at Roma Tre University, in Rome, and for other Master Programs at the University of Parma, Trani, Lecce and Matera. Since 2018, he has been serving as Scientific Director of the Higher Education Courses in Urban Strategic Design at the LUM School of Management in Bari. Since 2021, as part of this collaboration, he has been developing the guidelines for the urban regeneration of Italian cities for the SNA—National School of Administration.

Giovanni Leoni (1958) is Full Professor in History of Architecture at the University of Bologna where he directed the Department of Architecture, the Doctorate in Architecture and Design Cultures and was a member of the Board of Directors. His research focuses on the theory and practice of design between the nineteenth and twentieth centuries with particular attention to models of creativity, the relationship between Personality and Anonymity and the role of architectural design in social and political processes. He edits the journals *Histories of Postwar Architecture* and *European Journal of*

Creative Practices in Cities and Landscapes (with Carole Hein, TU Delft). He coordinated, for the University of Bologna, the H2020 project ROCK Regeneration and Optimisation of Cultural Heritage in Creative and Knowledge Cities and is part of the Bologna unit of the PNRR project CHANGES Cultural Heritage Active Innovation for Sustainable Society, Spoke 7, theme: The historic Italian city as a model for sustainable urban development. He has published extensively on the themes of architecture and the city in Emilia-Romagna in the second half of the twentieth century.

Contributors

Guya Bertelli Politecnico di Milano, Milan, Italy

Marko Bertogna University of Modena and Reggio Emilia, Modena, Italy

Andrea Borsari Alma Mater, Studiorum University of Bologna, Bologna, Italy

Dario Costi Smart City 4.0 Sustainable LAB, University of Parma, Parma, Italy

Federico Diodato Alma Mater Studiorum University of Bologna, Bologna, Italy

Sergio Duretti Lepida ScpA, Bologna, Italy

Andrea Fanfoni University of Parma, Parma, Italy

Gianluigi Ferrari University of Parma, Parma, Italy

Francesco Leali University of Modena and Reggio Emilia, Modena, Italy

Gabriele Lelli University of Ferrara, Ferrara, Italy

Giovanni Leoni Alma Mater, Studiorum University of Bologna, Bologna, Italy

Roberto Menozzi University of Parma, Parma, Italy

Carlo Alberto Nucci Alma Mater, Studiorum University of Bologna, Bologna, Italy

Emanuele Ortolan University of Parma, Parma, Italy

Francesco Pasquale University of Modena and Reggio Emilia, Modena, Italy

Michele Roda Politecnico di Milano, Milan, Italy

Smart City Myth and Challenge. An Open Laboratory to Promote a Debate Based on Six Key Concepts

Dario Costi

Abstract

This introductory chapter recounts the theoretical reflection that has been made on the theme of the contemporary city within the Smart City 4.0 Sustainable Lab. It also describes a multidisciplinary research method through the activation of a dialogue on four levels: disciplines, research, city and territory.

A few years ago, I had the pleasure of attending a lecture by the great classicist and author, Maurizio Bettini[1], who spoke at length about the role of Myth in ancient culture. He recalled the clash between two alternative dimensions of its meaning: the original definition of "authoritative discourse", as the reference of a great collective story that is handed down over time and resonates with each individual in a cultural and social community, and the later definition that downgraded the concept to "discredited narrative", loading it over time with negative reflections, exploitative uses and interpretative multiplications, reversing its significance to a sort of dispersion of meaning with different interpretations developed for partisan objectives.

From the strength of a civilisation that was reflected in shared discourses, to the weakness of a memory that has become negative, multiplied and distorted, the concept of Myth has come down to us delineating two opposing ways of thinking about and understanding collective themes. In the liquid society described by Bauman, there are no stronger myths than *Sustainability* and *Smart City*, two concepts we have been thinking about for some time and which inspired the name of the Research Laboratory we started in 2018.[2]

The first term is characterised in its first positive meaning, those of the origins and of Homer, increasingly as a shared authoritative Myth. The concept of *Sustainability* is becoming part of collective heritage, while widespread awareness is being activated concerning the environmental emergency and the outlook for the survival of the planet, thanks also to the initiatives of public institutions—from the community level to local authorities—and to actions linked to the implementation of UN Agenda 2030 by private actors,

[1] Maurizio Bettini was invited by Ivo Iori for a lecture promoted by the Fondo Librario Tassi of the University of Parma in October 2016. See the publication that includes this lecture in Maurizio Bettini, *Il Mito. Discorso autorevole o racconto screditato? (Myth. Authoritative discourse or discredited speech?)*, Tracce 6, il Mulino, Bologna, 2019.

[2] The *Smart City 4.0 sustainable LAB* research laboratory has been operational at the University of Parma since 2018 and at the other Universities of the Emilia-Romagna Region since 2019.

D. Costi (✉)
Smart City 4.0 Sustainable LAB,
University of Parma, Parma, Italy
e-mail: dario.costi@unipr.it

D. Costi and G. Leoni (eds.), *Smart City: A Critical Assessment*, The City Project 6,
https://doi.org/10.1007/978-3-031-51288-9_1

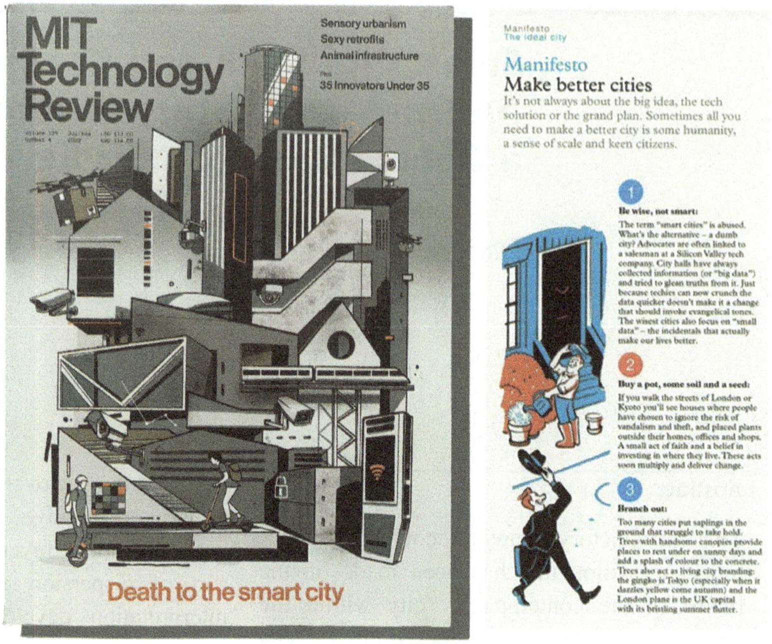

Fig. 1 On the left Volume 125 Issue 4 of the MIT Technology Review of July/August 2022 entitled *Death to the smart city*. On the right an excerpt from *The Monocle Guide to Building Better Cities*, Gestalten, Berlin, 2018, edited by Andrew Tuck where at the first point of the *Manifesto. Make better cities* is written: *"Be wise, not smart': "The term 'smart cities' is abused. What's the alternative—a dumb city?"*

who have been engaged for many years in a mobilisation that affects us all and involves many different sectors, from various types of associations to the most sensitive and responsible industries that have embraced the B Corp paradigm.[3]

The second term *Smart City*, on the other hand, remains suspended in an ambiguous dimension between the goal of becoming authoritative and the contingency of becoming a somewhat discredited fashion.

Everyone is talking about *Smart Cities* and many interpret the theme from partisan angles, describing experiences that may be of some interest but which elude an overall, integrated vision. There is still a lack of recognised scientific literature or convincing integrated examples, and the definitions that exist are either too broad and therefore vague or too specialised and therefore unrepresentative. From the start of the collective research experience, our Laboratory has always tackled the subject with a critical approach, which is reflected in a series of international publications, questioning all the generalisations being used to justify the content of a definition that emerges as dominant for the contemporary city (Fig. 1). The book that launched *The city project*[4] series, as well as this volume, are the first attempts at counterbalancing the tendency to identify the city as digital and fast with another city made up of tight-knit human relationships, brought closer together by recovering ancient values and approaches to rebuilding social relationships, using technologies not as an end but as a means. Innovation must be at the service of communities (Fig. 2).

Talking today about the contemporary city certainly means envisaging an intelligent city, but also a wise and sustainable one. Faced with the weaknesses brought to light by the 2020 pandemic, imagining this vision means keeping pace with the great impetus that the *Fourth Industrial Revolution* is bringing to our society and therefore also to the habits and rules with which we inhabit cities, thanks to a new dimension combined with the increasing virtual component. Within this historical transition, we believe that the future of cities must be constructed through a large-scale collective systemic action.

[3] In the context of Parma alone, some leading international companies such as Davines, Chiesi Farmaceutici and Barilla are driving this transition.

[4] https://link.springer.com/book/10.1007/978-3-030-76100-4.

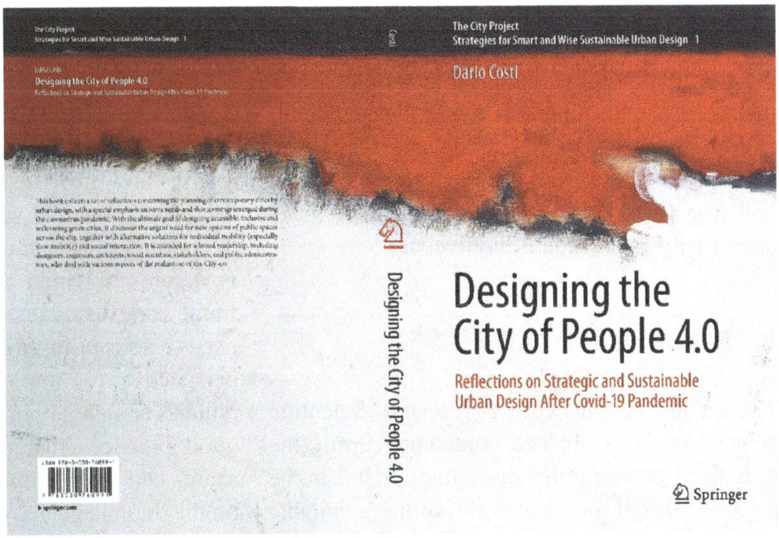

Fig. 2 Cover of the publication Dario Costi, *Designing the City of People 4.0, Reflections on Strategic and Sustainable Urban Design after Covid-19 pandemic*, Springer, Berlin, 2021

Therefore, the challenge of the contemporary city, through adhering to and at the same time broadening the concept (today perhaps still too instrumental, narrow and partial) of the *Smart City* lies precisely in developing a shared authoritative meaning, which does not yet exist, in the consciousness of the many stakeholders, thanks to the real-world experimentation of the operational possibilities that interdisciplinary and inter-institutional dialogue can develop within a clear cultural vision, which this publication seeks to present to the international scientific debate.

We think a collective and cross-party mobilisation must be activated to construct—from a largely processual perspective—a coherent meaning (yet to be explored) by prompting a dialogue based on six key concepts: *Disciplines, Research, City, Territory, Levels of Action and Working Directions.*

1 Disciplines

The first item in this dialogue is openness to all available expertise. Going beyond the strictly disciplinary dimension is in fact the first step. From the outset, the Laboratory that we decided to establish in order to address and discuss these issues set itself the goal of becoming an open intellectual container, a cultural space for exchange to bring together many of the scientific disciplines that have an interest in cities, starting with architecture and engineering. Thinking about the contribution each sector can make is the best way to go. We may need to transcend existing disciplinary boundaries, indicate some new directions and highlight the areas of expertise that can be activated in order to understand the potential of convergence on common ground for experimentation. We found it useful to start from possible applications rather than specific disciplines. A few examples may help to illustrate this variety: Industry 4.0, Internet of Things, dialoguing infrastructures, the role of water, circular economy, urban microfit, structural safety, sustainable planning, Urban Strategic Design, city history and cultural landscapes were the first contributions. However, many other fields of culture entered the arena too, contributing a number of different and complementary studies and perspectives. For example, disciplines that deal with nutrition, health and wellbeing, with highly relevant actions regarding training and social/health aspects; economic disciplines attentive to the social repercussions of production actions and to the processes of simplification and dematerialisation of management systems; legal disciplines that have been involved for many years in participatory policies, in proposals for framing international directives and responding to local needs for the use of heritage; disciplines such as sociology, anthropology, psychology, cultural geography, aesthetics; artistic disciplines such as music, cinema and art itself; and lastly, the sciences that deal with sustainability, with experience in the field of pollution control and green management as an opportunity to safeguard biodiversity.

2 Research

The second system word is research. Scientific debate is the obvious outcome from the encounter between the disciplines that can be activated. From the local scale, to the inevitable national and international projection, the university network is structured as a web of knowledge centred on the city that it will be crucial first to link up and then extend. The geographic homogeneity useful in finding specific solutions for the settlement conditions of each context—due to the obvious need to distinguish essential aspects such as urban history, settlement density, orography and climatic conditions in each situation—will need to be backed up by considerable openness to the international debate in order to broaden the horizon of possibilities to be applied and good practices to be adopted. Keeping the focus on the context and an attentive eye on the global scientific debate—within this interdisciplinary and inter-university perspective—is a double strategic action.

3 City

The third party players in this dialogue are those whose task it is to steer the transition to a smart and sustainable city. Authorities and public administrations must be the partners in this overall systemic action from the outset. The sharing of themes and the identification of targeted pilot projects that can introduce the various skills and the necessary technologies within scenarios outlined by the urban project developed with a strategic and participatory approach become the empirical verification condition for experimentation that provides real opportunities to get the skills involved to interact profitably. Institutional stakeholders identified the goal of implementing the *Smart City* some time ago but, in all honesty, in many contexts they do not really know how to achieve it, while in others they apply a partial version of it normally linked to contingencies and opportunities. Research that is organised within an integrated and dialogic framework needs institutions, just as institutions need research in order to provide an overall perspective for implementation. Many sectoral policies are already in place and can contribute to good practices that public administrations can share. One could think, for example, of actions on the digital agenda, experiments in smart urban areas, social digital innovation, or environmental and sustainable mobility policies, where the mere involvement of institutional bodies is not enough to produce the desired results.

4 Territory

Each context must initiate an open dialogue with civil society in its many forms, involving study centres and foundations, trade associations, the world of cooperation, public health agencies, institutional stakeholders dealing with landscape and territory, professional associations and other local entities based on an open-source approach that maximises sharing. Interaction with all available and interested parties is the basis for a useful dialogue to share and add value to the initiatives already in place and to highlight ideas to be developed in order to respond to emerging needs in the various fields of human action undergoing rapid change. Talking to all the stakeholders throughout the territory was from the outset a prerequisite for expanding a transparent and open network of dialogue as far as possible, where all the nodes are constantly prompted, informed and called upon for their expertise. A public, open and inclusive territorial pact between research, public administrations and stakeholders is the ideal framework for practical implementation of theoretical reflections on the contemporary city, in any context.

Fig. 3 The *Smart City 4.0 Sustainable LAB* network

5 Levels of Action

As of 26 March 2018, with the institutional presentation at the University of Parma, the *Smart City 4.0 Sustainable LAB* research laboratory began its work to support public bodies, institutions, and stakeholders in the territory. A series of preliminary meetings with the municipalities of Modena, Reggio, Parma and Piacenza identified the needs of the cities of Emilia-Romagna, which led to the development of a series of themes to be addressed.

The subsequent establishment of a coordination of professors from the Universities of Piacenza (Cattolica and Polytechnic), Reggio Emilia and Modena, Bologna and Ferrara, as well as Parma, provided a basis for sharing strategies and timeframes. The expansion of the Laboratory on a regional basis brought together lecturers and researchers in the relevant disciplines and innovative technologies, resulting in a broad-based shared project. The Research Laboratory thus established a network of researchers spread across the region in a four-tier structure (Fig. 3).

The **Regional Scientific Committee**, of international standing, is made up of professors and researchers from local universities, experts invited from all over the world, representatives of the Emilia-Romagna Region and a representative of Lepida ScpA, a company involved in the design and testing of Information and Communication Technology products.

The **Territorial Table for the Contemporary City of Emilia-Romagna** is a federation of experts on the themes of the city and the territory with the aim of tackling the challenges of the European context and building together a framework for training and for structuring of actions that is as extensive as possible. The Territorial Table sees the involvement of the Federations of Professional Orders of Engineers, Architects, Surveyors and Experts, ANCE (National Association of Building Constructors), ANCI (National Association of Italian Municipalities), CNA Emilia-Romagna (National Confederation of Craftsmen and Small and Medium Enterprises), Legacoop Emilia-Romagna, Arpae (Regional Agency for Prevention, Environment and Energy of Emilia-Romagna), ADBPO (Po River District Basin Authority), Regional Agency for Territorial Safety, Civil Protection and AIPo (Interregional Agency for the Po River).

The **interdisciplinary/inter-university network**, made up of research groups distributed among the various universities involved, forms the scientific basis of the Laboratory and is divided among all the disciplines that can contribute to the innovation and enhancement of the contemporary city.

The **interdisciplinary/inter-university working group** consists of Ph.D. students, supervisors and assistant supervisors from the universities in the region. Since mid-2019, the Research Laboratory has seen the establishment—through the call for proposals *High Skill Projects for the Smart and Sustainable Contemporary City POR FSE 2014/2020 Thematic Objective 10*—of five Ph.D. scholarships supported by the Emilia-Romagna Region in the various university locations. The scholarships are integrated in an inter-university working group that aims to produce specialisations on Smart City and Smart Land through a permanent and linked action.

6 Working Directions

Since 20 May 2019, the date on which a memorandum of understanding was signed involving the University of Parma, the Emilia-Romagna Region and Lepida ScpA, the Laboratory has been committed to carrying out its research work by following eight working directions/missions (Fig. 4). These have been accomplished thanks to the support of institutional partners and to operational coordination involving Universities, Bodies and Institutions in the territory. This enabled the establishment, as of December 2021, of two high-level training courses and two annual Summer Schools financed by the Region, which have transformed theoretical and cultural reflections into practical and operational demonstrations. The eight working directions, are still valid for the future, are:

1. **University research aimed at defining a working methodology for Public Administrations** guides actions for the implementation of the Smart City through applications in case studies proposed by municipalities and local authorities.
2. **Pilot Projects for Municipalities** that, with the support of the Emilia-Romagna Region, have been called upon to identify emerging themes in their context, which serve as useful case studies for the drafting of general guidelines.
3. **The Pilot Projects for local authorities and stakeholders** who have agreed with the Laboratory on the themes they would like to be addressed both to improve their services and activities and to derive general guidelines that have been collected in the intervention methodology.
4. **Training of Public Administrations and identification of themes** within schools of Higher Education on the Smart City have proved to be a valuable tool to support dialogue between public decision-makers and universities and to identify critical points in urban contexts on which to intervene.
5. **The relationship between research and companies.** Companies and economic operators active in the cities have been linked to the Research Laboratory, agreeing on joint working directions and synergistic actions.
6. **Specialisation on the part of professionals, technicians and operators** within the territory was achieved through the involvement of professional associations, regional Federations of Professional Orders, CNA (National Confederation of Craftsmen and Small and Medium Enterprises) and Confindustria (Italian Manufacturing Companies Association). These had the opportunity to update their skills by supporting the work of the Research Laboratory and participating in advanced training courses on the Smart City directly linked to the other parallel actions.
7. **The territorial federation of universities, institutions, and enterprises for participation in European calls for proposals** was created with the aim of establishing an organised network to compete and dialogue on an international level.
8. **Documentation, dissemination and involvement** were complemented by dedicated programmes of participation and awareness-raising among the settled communities. This led to the organisation of international conferences, the publication of research activities and a widespread growth of awareness regarding the collective challenges of the contemporary smart and sustainable city.

Fig. 4 The eight working directions of the *Smart City 4.0 Sustainable LAB*

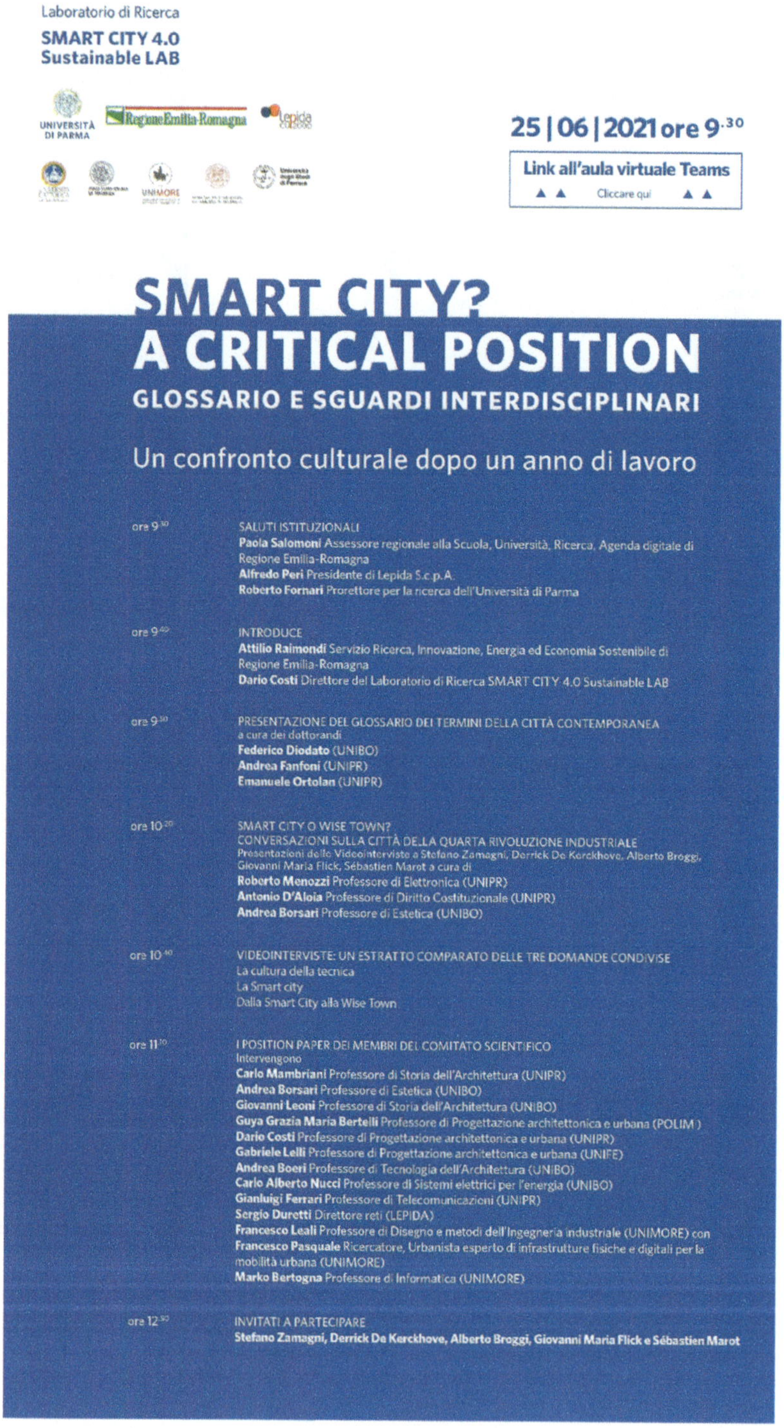

Fig. 5 Poster for the conference *Smart City? A critical position. Glossary and interdisciplinary perspectives. A cultural exchange after a year of work* (25 June 2021) where the Glossary of Contemporary City Terms and the video interviews that are published in this volume were presented

Thanks to this type of work in 2022, we can confirm that we have fulfilled the commitments made at the beginning of the Laboratory's journey with the Emilia-Romagna Region. On a daily basis, we verify the need for this dialogue based on these six key concepts. Advancement of knowledge and perspectives will only be possible if everyone is committed to taking on a great collective challenge. This is the only way we will be able to work together to devise a model for cities and their territories, on a case-by-case basis, that makes the most of the skills and experience currently available. Only in this way can we restore the *Smart City* to its useful and more than ever necessary dimension as a positive myth, because it is shared, as an authoritative narrative of our communities.

7 Publication Context

This publication is the outcome of the findings that emerged from a conference planned during the Covid-19 lockdown to discuss, develop and produce a critical scientific position on the concept of the Smart City, challenging an unqualified and all too widespread use of this concept that is often equated in meaning with the concept of the digital city, far removed from our approach of integrating scientific and humanistic culture and imagining a dialogue between disciplines (Fig. 5). This publication is the synthesis of a discussion that was then in the set-up phase and that today retains proven value in the research activities and actions pursued by the Laboratory's researchers from 2018 to 2022.

The following volume is divided into three parts. The first part is a Glossary compiled through a review of the latest developments, definitions, terms and concepts revolving around the contemporary city. The concept of *Smart City* can be very vague today. There are indeed many interesting experiences and practices, but a synthetic definition and a clear course of action are perhaps still lacking. The Ph.D. students of the Laboratory conducted a very interesting survey of a number of key words that demonstrates the need to define certain fundamental concepts.

The second part concerns a dissemination initiative, organised by the University of Parma, which we saw as an opportunity for research. These are video interviews in the form of dialogues between a very diverse selection of speakers in terms of culture, action and thought on the themes of the contemporary city. We asked ourselves who could help us to focus on the problems of the city, identifying authoritative speakers who represent national and international cultural references in their various disciplines. Not just architects, engineers and urban planners, but professionals who study the city from many different points of view: an economist, a philosopher, an autonomous driving engineer, a media sociologist and a constitutional lawyer. The first three questions put to the experts are the same. This made it possible to have a comparable discussion, which elicited deeply thought-provoking views on the city.[5]

The third and final part consists of eleven position papers written by members of the laboratory Scientific Committee who were asked to comment on the significance of their research. These papers provide a starting point for a sustainable development capable of bringing together all disciplines that deal with cities.

[5] The video interviews can be viewed at the following links:
Stefano Zamagni (https://www.youtube.com/watch?v=eAkUx9heQYw).
Derrick De Kerckhove(https://www.youtube.com/watch?v=VlBVOhKQaKs).
Alberto Broggi (https://www.youtube.com/watch?v=kCi6jR3aksE).
Giovanni Maria Flick (https://www.youtube.com/watch?v=P3UAUH51zNg).
Sébastien Marot (https://www.youtube.com/watch?v=jc1jvChwDqw).

Dario Costi, Arch. Ph.D., is Full Professor in Architectural and Urban Design in the Department of Engineering and Architecture at the University of Parma. He is the Director of the interdisciplinary teaching and research project "Smart City 4.0 Sustainable LAB", which is promoted by the University of Parma in cooperation with the Emilia-Romagna regional government and the public digital infrastructure Lepida, and involves the Universities of Ferrara, Bologna, Modena and Reggio-Emila, Piacenza as well as local Authorities and stakeholders. He is member of the Ph.D. Program in Civil Engineering and Architecture at the University of Parma and at the Faculties of Architecture of the University of Palermo and Sapienza University of Rome. He is active both as a professional and as a researcher in the architectural and urban design field. He is the co-owner of the MC2 Architecture Studio in Parma. Since 2007 he is serving as President of Parma Urban Center, of which he is also the co-founder. Involved in numerous national projects for fostering education in architecture and urban design, Professor Costi is the author of many Italian publications, and Editor of two Italian book series published by MUP Editore, Parma. Since 2020, he is the editor of the international book series "The City Project—Strategies for Smart and Wise Sustainable Urban Design", published by Springer. He is also serving as a lecturer for the Program Master Housing at Roma Tre University, in Rome, and for other Master Programs at the University of Parma, Trani, Lecce and Matera. Since 2018 he is serving as Scientific Director of the Higher Education Courses in Urban Strategic Design at the LUM School of Management in Bari. Since 2021, as part of this collaboration, he has been developing the guidelines for the urban regeneration of Italian cities for the SNA—National School of Administration.

Glossary of Contemporary City Terms. A Critical Selection of Definitions in the Literature Towards the City of People 4.0

Federico Diodato, Andrea Fanfoni, and Emanuele Ortolan

Abstract

The drafting of a *Glossary of Terms for the Contemporary City* stems from the need, perceived by the Scientific Committee of the *Smart City 4.0 Sustainable Lab*, to collect, catalogue and critically analyse the vast semantic spectrum of concepts and terms used in the literature to describe city development, organisation and control models from the start of the 21st century to the present day.

1 Reasons for a Glossary

The drafting of a *Glossary of Terms for the Contemporary City* stems from the need, perceived by the Scientific Committee of the *Smart City 4.0 Sustainable Lab*, to collect, catalogue and critically analyse the vast semantic spectrum of concepts and terms used in the literature to describe city development, organisation and control models from the start of the 21st century to the present day.

The *Glossary* is divided into three parts.

The first part identifies background terminology useful for contextualising and delimiting the scope of research to the city of the *Fourth Industrial Revolution*, focusing on the role of technology in urban transformations and on the complex relationship between technology and the dynamics of territorial government.

The second part collects, breaks down and examines a number of definitions in the literature relating to four neologisms that identify four different but sometimes overlapping urban models: the *Digital City*, the *Smart City*, the *Green City* and the *Wise City*. For each term, an exhaustive sample of definitions were catalogued, the key attributes extrapolated from the statements analysed were identified, and the resulting synthesis led to a definition proposed by the laboratory's PhD students.

The third and final part clarifies the purpose of the critical analysis carried out on the models observed by envisaging a development scenario aimed at integrating the city of technological opportunities with the city of individuals: the *city of people 4.0*.

The current crisis situation (environmental, economic, health) has made new technologies crucial to the functioning of cities. They have been the best response to the increasingly urgent need for automation and organisation of professional and social life. However, this situation

F. Diodato
Alma Mater, Studiorum University of Bologna, Bologna, Italy
e-mail: federico.diodato2@unibo.it

A. Fanfoni (✉) · E. Ortolan
University of Parma, Parma, Italy
e-mail: andrea.fanfoni@unipr.it

E. Ortolan
e-mail: emanuele.ortolan@unipr.it

D. Costi and G. Leoni (eds.), *Smart City: A Critical Assessment*, The City Project 6,
https://doi.org/10.1007/978-3-031-51288-9_2

also exposed the limits of a technicist approach and the growing concerns around control devices that seem to be establishing an increasingly pervasive relationship with the people who use them. The relationship between new technologies, cities and territory is a complex one, on which we must reflect in order to define planning strategies that can contribute to "*greater social equality and sustainability in the recovery*".[1]

To guide us through this complexity, and to envisage "an intelligent city" that is not just a neo-technologically based infrastructure, the Glossary provides a critical definition of some of the concepts that, since the earliest studies on cybernetics and artificial intelligence in the 1950s, have been propounding the possibility of a form of non-human intelligence that could contribute to the functioning of the urban system.

With the development of Information and Communication Technologies (ICT), we have seen a radical shift in the way we relate to technology, participating in the development of a "*technological society*", i.e. a society in which technological progress is predominant.

We now live in "*environments*",[2] surrounded on all sides by devices that organise and determine social constructions, which, through collecting and processing *big data*, structure "*the whole informational environment constituted by all informational entities, their properties, interactions, processes, and mutual relations*".[3] This information environment not just confined to its digital dimension. It is organised in a constant relationship between online and offline, in which the boundaries between physical and digital are increasingly blurred to the point of disappearing. The need to collect an ever-increasing amount of data has meant that the industry dedicated to online storage systems, *cloud computing*, has undergone enormous development in recent years.[4] However, cloud storage systems require a physical infrastructure, *data centres*, which host a set of digital infrastructures necessary for data storage and processing.

Although the physical impact of the infrastructures required for the systems and their operation is studied increasingly in the scientific literature,[5] the narratives that propose the use of new technologies for urban and territorial planning tend to underestimate this impact, reducing the territory to its informational dimension alone, disconnected from its physical, social and cultural complexities, and considering the reality of the territory as abstract and timeless.[6] The risk is that of considering intelligence as the prerogative of technological innovations, without considering that the territory is the outcome of a complex relationship between social models, economic relations and the physical configuration of space.

The Glossary is an attempt to offer a critical definition of the most relevant terms, which fully considers the aforementioned complexity so as to "*direct innovation towards the most authentic human needs*", allowing technological development to "*be complementary to human evolution and have an emancipatory function, to liberate and shape the time and space available to mankind*".[7]

The first term analysed in the glossary is *Digital City*. The concept is deeply rooted in the thinking of sociologist Marshal McLuhan, who as early as 1964 spoke of a "*global village*"[8], where the social and communicative dimension of the village, characterised by relationships of proximity, was extended to the entire planet thanks to a merging of space and function made possible by the acceleration of the electronic age and human interactions filtered by the technological medium.

In an article written in 1990 entitled "*The computer for the 21st Century*", Mark Weiser,

[1] Klaus SCHWAB, Thierry MALLERET, *COVID-19: The Great Reset*, Geneva, Forum Publishing, 2020, p. 377.

[2] Luciano FLORIDI (Eds.), *The Onlife Manifesto*, Cham, Springer International Publishing, 2015.

[3] Luciano FLORIDI, *The Fourth Revolution: How the Infosphere is Reshaping Human Reality*, Oxford, OUP Oxford, 2014, p. 41.

[4] Luciano FLORIDI, *The Fourth Revolution, op. cit.*

[5] Fanny LOPEZ, *L' Ordre électrique: Infrastructures énergétiques et territoire*, Paris, MētisPresses, 2019, p. 208

[6] Alberto MAGNAGHI, *Il progetto locale*, Torino, Bollati Boringhieri, 2010, 344 p.

[7] Paolo ROSA, Andrea BALZOLA, *L'arte fuori di sé*, Milano, Feltrinelli, 2011.

[8] Marshal MCLUHAN, *Understanding Media: The Extensions of Man*, McGraw-Hill, New York, 1964.

computer scientist and director of the Xerox investigation centre, foresaw the use of ubiquitous and invisible information technology that would have to disappear from sight in order for it to be accepted by human psychology.[9] Starting from these assumptions, sociologist Manuel Castells states that the modern city is characterised by increasing articulation and simultaneous structuring and deconstruction between physical space and the space of flows, where the former organises experience within the limits of geographical location, while the latter establishes connections between physically separate places.[10] Castell argues that the combination of place and network would nullify the notion of time, realising the paradox of life in the urban environment even in the absence of the city. For this reason, this situation will have to be remedied by recovering the symbolic relevance of public space, restoring the city to its ancient function through interventions to conserve places that have a collective purpose, emblematic of the urban experience.

The definitions analysed in the glossary identify the *Digital City* as an "*urban relationship space*" augmented by the implementation of information technologies that facilitate communication through virtual means. In general, the definitions identify elements in the model that enable an improvement in the quality of life, coexistence between citizens, security and communication between inhabitants and institutions. Other definitions highlight the increased possibilities for dialogue facilitated by simulated public spaces for sharing experiences that promote the activation of cooperation strategies for the well being of the entire community. In addition, the potential generated by the creation of virtual metaphors of the real city, accessible to all and usable by public administrations to improve physical spaces through careful analysis of data collected by sensor networks, is emphasised.

In general, the definitions analysed do not contain references—except in passing—to the risks that McLuhan foresaw as early as the 1960s, such as the loss of notions of space and time with the consequent loss of capacity for assuming social roles and individual points of view, or the need to recover the spatial forms typical of urban experience as advocated by Castells. The fact that ubiquitous technology acts on the social dimension of the subject can be deduced from a variety of evidence,[11] but how technology conditions the physical dimension of the city is a field of study yet to be explored.

The questions that the authors pose are how can the *Digital City* facilitate relations between citizens without undermining the city as a socio-spatial communication system? How can we avoid the loss of awareness resulting from the implementation of technology that may have ethical implications in terms of forcibly controlling the lives of the city's inhabitants?

The answer to this complex set of questions is addressed in the Authors' definition, where they highlight the need for the *Digital City* to become the platform where physical space and virtual space meet in a context of mutual cooperation, while warning about the risks that may arise from the loss of the perception of reality and reminding us that the ultimate goals resulting from the urban use of technologies are to increase the accessibility of all citizens to community life by enabling and facilitating individual participation in the connected city, ultimately improving mobility and logistics, security, resilience to adverse environmental factors and the sustainability of the urban system.

The second term analysed in the glossary is *Smart City*. Among the 29 definitions observed, an urban model emerges focused on the growth of the quantity of services through the massive use of embedded smart device technology: the *Internet of Things (IoT)*. It is also clear that the definitions highlight a development strategy

[9] Mark WEISER, *The computer for the 21st Century*, Scientific American, September, 1991, pp. 94–104

[10] Manuel CASTELLS, *La città delle reti (The Network City)*, Marsilio, Venezia, 2004.

[11] Francisco Javier ANSUÁTEGUI ROIG, *Nuove tecnologie e spazio pubblico (New technologies and public space)*, in Silvia SALARDI, Michele SAPORITI (Eds.), *Le tecnologie 'morali' emergenti e le sfide etico-giuridiche delle nuove soggettività (Emerging 'moral' technologies and the ethical-legal challenges of new subjectivities)*, Giappichelli Editore, Turin, 2020.

based on the six dimensions introduced in 2007 by the Vienna *Center of Regional Science*: *Smart Governance*, *Smart Living*, *Smart People*, *Smart Mobility*, *Smart Economy* and *Smart Environment*.[12] Accessibility to services, efficient public administration, policies for inclusiveness, cooperation with knowledge-intensive activities, support for creativity and the growth of human capital, control of road infrastructure, economic competitiveness of the labour market, actions in favour of environmental protection, energy efficiency and the reduction of pollutant emissions are just some of the characteristics that identify a *Smart City* in the definitions collected.

On the other hand, only a handful of definitions focus on the quality of the physical environment and on urban design as a development strategy. Evidence of the latter led the authors to question the efficiency of Smart proposals through an open multidisciplinary debate.

An analysis of the Glossary's contents allows certain critical aspects of the model to be identified, such as the lack of evidence that ICT technologies actually improve the democratic process, the danger of a progressive replacement of human functions by machine functions, the possibility that technology will not help to make the city more inclusive unless it is guided by a public purpose, the possibility that constant connectivity will lead to difficulties in the area of individual freedoms and personal autonomy, the creation of inequalities caused by the digital divide and the unequal distribution of technology, the disappearance of institutional intermediaries, the invasion of privacy brought about by Big Data through the misuse and illegitimate use of the data provided, and, as already mentioned, the lack of a qualitative assessment of public space and urban development planning processes. This series of topics is intended as the critical contribution of the publication based on the definitions collected and forms the basis of the debate generated among the experts and the members of the Scientific Committee of the Laboratory who participated.

It is also interesting to note how the challenges posed by climate change, the need to reduce pollutant emissions and, lastly, the pandemic caused by Covid-19 have highlighted the need to work not only on technological aspects, but also on environmental aspects linked to the urban context, emphasising the need to harness technology to improve the quality of the natural environment that characterises a city and a territory. We are therefore moving away from the *Smart* city concept towards *Green* models, where environmental issues are increasingly prioritised.

The English term *Green* is used, in its broadest sense, to refer to issues related to the protection of the natural environment and to imbue an action or activity with a connotation that embodies the principles of environmental sustainability, resulting in the concepts of *Green City*, *Garden City* or *Sustainable City*. This is a city model tried and tested at European and international level that focuses on high environmental quality and biodiversity, not as an isolated and delimited objective, but as part of a broad urban regeneration design for the benefit of communities, with a focus on two main dimensions: environmental and social.[13]

The first dimension, through urban regeneration, has the twofold objective of making cities more attractive by developing healthy and quality urban environments, regenerating marginal spaces, while minimising environmental costs through the progressive reduction of land consumption and the decarbonisation of urban systems, with clear benefits in terms of resilience in the face of climate change. The second dimension —the social one—which clearly emerges from the definitions surveyed, emphasises the role of human beings in the use and organisation of natural resources in order to prevent environmental pollution and ensure the mindful use of

[12] Rudolf GIFFINGER, Christin FERTNER, Hans KRAMAR, Robert KALASEK, Nataša PICHLER-MILANOVIC, Evert MEIJERS, *Smart Cities: Ranking of European medium-size cities*, Center of Regional Science, Vienna University of Technology, 2007.

[13] This multidimensional approach based on environmental and social aspects, with clear health and economic benefits, is promoted internationally by the UN Environment Programme (UNEP), the Organisation for Economic Co-operation and Development (OECD) and the European Landscape Contractors Association (ELCA).

natural resources. For the first time in urban models, the need to promote responsible actions and virtuous behaviour is emerging, which in turn contributes to positive economic spin-offs for the achievement of the goals set by a "green city".

In recent years, it has been observed how the terms *Smart* and *Green* have been defined and framed within the urban sphere and have entered the common lexicon of administrators, public decision makers, planners, private stakeholders and citizens. There are dozens of definitions in the scientific literature: in some, environmental protection is more relevant, in others it is digitalisation, in others it is the infrastructure endowment of cities, while in others it is the social dimension. At the same time, the need has emerged to broaden the meaning of both *Smart City* and *Green City* and envision a system to enhance the habits of city dwellers with the new opportunities for interconnection and digitalisation. In addition to the speed and intelligence of the *Smart City* and the environmental qualities attainable through *Green* models, we can also imagine a *Wise City*, that is, a wise and reflective city reorganised around people and social relationships.[14]

There is still no generally accepted definition of *Wise City* in the scientific literature. In order to find one that fits the hypothesised imaginary, one must go back to the definition of *Smart City* and distributed intelligence, incorporated in the public space, as a redefinition of the ancient *genius loci*, demonstrating that the debate for a transition to a wise and sustainable city for communities is more open than ever.

Indeed, the etymology and meaning of the adjective *wise* highlight some fundamental specificities that distinguish the *Wise City* from the *Smart City*. This distinction can also be deduced from certain attributes and terms that define the adjective *wise* as sensitive, sensible, balanced, judicious and knowledgeable, insisting on the aspect of awareness. The transition from *Smart City* to *Wise City* will therefore be the result of a conscious choice in defining, through planning, a new urban identity. By translating it into its urban dimension, the concept of the *Wise City* and a number of its synonyms, including the *Intelligent City*, *Knowledge City* and *Care City*, can then build on the technical aspects that characterise the Smart City. In this case, however, the technology and the benefits it brings will support the citizen, who becomes a protagonist in shared choices and no longer merely the recipient of technological development. The *Wise City* can therefore be configured as an evolution of the *Smart* concept, where the huge emphasis on smart devices and their ability to gather information can be harnessed to serve a compact urban development, which can be pursued through strategies of increased density, typological variety, enhancement of public space and rationalisation of infrastructures, now reusable as a matrix of social experience, for a paradigm shift in urban mobility.

However, it is important to pay attention to the possible limits of a further model that risks, like the previous ones, to become easily scalable and adaptable to be imposed without distinction on any context for the transformation of urban centres and territories.

Instead, the idea of a Wise City, or rather a Wise Town, a term that refers to a more human measure, may represent the starting point to return to thinking about inhabited centres as a place of organised coexistence and the city as a common property promoted by interested citizens, introducing new thinking around dimensions such as balance, prudence and the proportioning of things.

2 Preliminary Terminology

Cybernetics
From the Greek κυβερνήτης, (implied τέχνη) 'the art of piloting', from κυβερνάω 'I steer a ship'.

[14] Robin HAMBLETON, *From smart cities to wise cities*, Paper for the AAG Annual Meeting, Chicago, Illinois, 21–26, April 2015.

Definition	Source
Discipline that studies the similarities between the regulation and communication systems of machines and living organisms, and in particular the application of natural regulation mechanisms to technology	Translated from Italian *Dizionario lo Zingarelli*, 2021
A discipline that deals with the unitary study of processes concerning "*communication and control in the animal and in the machine*" (according to the definition of N. Wiener, 1947): starting from the hypothesis that there is a substantial analogy between the 'regulation mechanisms' of machines and those of living beings and that at the basis of these mechanisms there are communication and information analysis processes, the c. proposes, on the one hand, to study and construct machines with a high degree of automatism, capable of replacing man in his role as controller and pilot of machines and systems, and on the other hand, conversely, to use the aforementioned machines to study certain physiological and intelligence functions. (...)	Translated from Italian *Enciclopedia Treccani online*

Definition Proposed by the PhD Students of the Smart City 4.0 Sustainable Lab

Cybernetics studies in the 1950s and 1960s and early forms of artificial intelligence played a predominant role in the development of smart planning models,[15] contributing to the spread of one of the founding myths of the digital age: the possibility of a non-human form of intelligence that could make a functional contribution to the functioning of the urban system. Antoine Picon observes that in the cybernetic approach, the city is considered as an information system based on "*une vision résolument réductionniste de l'homme et du monde*".[16] The development of cybernetics belongs to the field of cognitive studies that have thought of the brain as an advanced information system. Norbert Wiener, the forerunner of cybernetics, already warned at the end of the 1940s about the risks that the power of cybernetic innovations could entail for the control of the human being: if technology were to fulfil our desires to the letter to bring us benefits, are we sure that we are asking for what we really want and not what we think we want?[17]

Cloud

From the English language

Definition	Source
In computer science, cloud computing; specifically, the set of hardware or software resources on remote servers and distributed over a network, containing a user's data and programmes (...).	Translated from Italian *Enciclopedia Treccani online*

Definition Proposed by the PhD Students of the Smart City 4.0 Sustainable Lab

The need to collect an ever-increasing amount of data has caused the industry dedicated to online storage systems, cloud computing, to undergo exceptional development in recent years.[18] Essential technologies for the development of smart planning models, cloud storage systems require a physical infrastructure known as data centres, facilities that house a set of digital infrastructures necessary for storing and processing data. Researcher Fanny Lopez indicates that smart city studies tend to disregard the environmental impact of the infrastructure required for cloud storage systems, leaving out the role that data centres play in electricity consumption, stating: "*les data centres seront parmi les plus importants postes de consommation électrique du XXIe siècle*".[19] How to develop data centres that have a reduced environmental impact and are efficient in electricity consumption is one of the challenges that a smart smart planning project must be able to respond to.

Infosphere

(information + -sphere) analogous to a biosphere, cryosphere, etc.

Definition	Source
The virtual space consisting of all the information circulating in the world by both traditional means (press, radio and television) and digital means (Internet or other networks)	Translated from Italian *Dizionario lo Zingarelli*, 2021
The set of media and the information that is produced by those media. It is a society	Translated from Italian

(continued)

[15] Antoine PICON, *Smart Cities: Théorie et critique d'un idéal auto-réalisateur*, Editions B2, Paris, 2013.

[16] Ibidem.

[17] Norbert WIENER, *Cybernetics or Control and Communication in the Animal*, Mass: MIT Press, Cambridge, 1948.

[18] Luciano FLORIDI, *The Fourth Revolution: How the Infosphere is Reshaping Human Reality*, OUP Oxford, Oxford, 2014.

[19] Fanny LOPEZ, L' Ordre électrique: Infrastructures énergétiques et territoire, MētisPresses, Paris, 2019.

Definition	Source
dominated by technologies that make possible, as of today, a profound change in the 'infosphere', thanks to the demassification of the media. (...)	*Enciclopedia Treccani online*
Definition Proposed by the PhD Students of the Smart City 4.0 Sustainable Lab	
The development of increasingly pervasive information and communication technologies constitutes an 'infosphere'. A concept formulated in the 1980s from the concept of the 'biosphere', the 'infosphere' is properly the information environment and is defined by the philosopher Luciano Floridi as "*the whole informational environment constituted by all informational entities, their properties, interactions, processes, and mutual relations*".[20] Extrapolating the concept, one can consider the 'infosphere' as synonymous with reality itself interpreted as an information relationship. From a spatial point of view, its main peculiarity is its in-between state: in a constant relationship between online and offline, the boundaries between physical and digital are increasingly blurred until they disappear. Such is the condition of 'onlife',[21] the vital dimension in which human beings conceive of themselves as 'inforgs'. By 'inforgs', Floridi specifies, it is not a question of a biotechnological transformation of the human being. biotechnological transformation of the human being, rather it is the environment itself that transforms and adapts.[22]	

Technique

From the Greek τέχνη art and propr. ability to produce.

Definition	Source
1. Set of rules governing the concrete performance of a manual, intellectual or sporting activity 2. Way of working, producing, realising something 3. Any form of human activity aimed, by exploiting the knowledge and acquisitions of science, at the creation of new means, instruments, devices, apparatus that improve the conditions of life of man himself	Translated from Italian *Dizionario lo Zingarelli*, 2021
In an abstract and generic sense, the set of practical activities based on empirically acquired norms, or on tradition, or on the application of scientific knowledge, characteristic of a given social and productive situation, of a given epoch, of a given geographical area: subject to historical evolution, it is characterised by a set of reciprocal relations (...). Technology is configured as a heritage of knowledge, increasingly specialised and subject to continuous innovation, requiring specific specific training. (...)	Translated from Italian *Enciclopedia Treccani online*
Definition Proposed by the PhD Students of the Smart City 4.0 Sustainable Lab	
According to the philosopher Adriano Fabris, technique would be 'born' from the awareness of a lack, from the consciousness of one's own limits and shortcomings'[23] An intermediary between theory and practice, technique would turn out to be a form of application of scientific knowledge to the practical realm of reality, endowing human beings with the tools (technical instrument) to fill this gap. The technical tool is a functional object that, inserted into a system of relations, becomes the means to achieve it. In this sense, the technical tool allows the living being to extend its possibilities of action and to adapt to its environment. To achieve this common goal, the anthropologist Leroi-Gourhan points out, the technical tool put into a system externalises the sensory functions of the human being[24] In this process of 'externalisation', the technical tool is characterised by its removability and substantial distinction from the bodies themselves: while the technique invents man, man invents himself in the technique by inventing the tool.[25]	

(continued)

[20] Luciano FLORIDI, *The Fourth Revolution: How the Infosphere is Reshaping Human Reality*, OUP Oxford, 2014), p. 41.

[21] Luciano FLORIDI (Eds.), *The Onlife Manifesto*, Springer International Publishing, Cham, 2015.

[22] Luciano FLORIDI, *The Fourth Revolution: How the Infosphere is Reshaping Human Reality*, Op. Cit.

[23] Adriano FABRIS, *Etica per le tecnologie dell'informazione e della comunicazione*, Carocci, Roma, 2018, p. 13.

[24] André LEROI-GOURHAN, *Le geste et la parole, Tome 1: Technique et langage, Sciences d'aujourd'hui*, Albin Michel, Paris, 1975.

[25] Bernard STIEGLER, *La technique et le temps 1. La faute d'Épiméthée*, Galilée la Cité des sciences et de l'industrie, Paris, 1994.

Technology

From the Greek τεχνολογία, compound τέχνη art and λόγος speech, treatise.

Definition	Source
Study of the technique and its application	Translated from Italian *Dizionario lo Zingarelli*, 2021
Multidisciplinary research field with the subject of the development and application of technical tools, i.e. what is applicable to the solution of practical problems, the optimisation of procedures, decision-making, the choice of strategies aimed at objective data, on the basis of scientific knowledge including mathematical and computer science	Translated from Italian *Enciclopedia Treccani online*

Definition Proposed by the PhD Students of the Smart City 4.0 Sustainable Lab
Technology is understood as the systemisation of technical tools for the achievement of a common goal. At present, technology is not just an externalisation of the human being's senses, it is an extension of the central nervous system, which would imply an increase in cognitive performance.[26] In this sense, the technological device is pervasive and differs from the technical tool by its progressive emancipation from the human being: technology is no longer the product of human initiative, but is freed from human control.[27] The rise of digital technologies has developed a 'technological society', that is, a society in which technological progress is predominant and in which it is impossible to distinguish the technical tool from the user. "Environments" in which the human being is surrounded by ubiquitous devices that organise and determine social constructions.

ICT

Acronym for Information and Communication Technology.

Definition	Source
Abbreviation for information and communication technology: the use of computers and other electronic equipment and systems to collect, store, use, and send data electronically. ICT can improve transparency and accountability in government and private sector operations	*Cambridge Dictionary*
Acronym for information and communication technologies, a term denoting all means of calculation, information and communication as well as the study of the potential of their integration, also increased by the development of the Internet. The acronym indicates the current development of that branch of the discipline, on the borderline between mathematics and computer science, originally called automatic calculation	Translated from Italian *Enciclopedia Treccani online*

Definition Proposed by the PhD Students of the Smart City 4.0 Sustainable Lab
Information and communication technologies are generally concerned with the processing and transformation of data that are transmitted in information processes. These are devices whose evolution has radically changed the way we relate to the machine: we cannot just act on them but must constantly interact with them. Their use is crucial in enabling the optimisation of urban services promised by Smart models. In the ITC acronym, Adriano Fabris points out,[28] the undifferentiated identification of the processes of 'information' and 'communication' would make the communication process lose its meaning of involvement and active participation, reducing it to the simple process of transmitting a message

[26] Marshall MCLUHAN, *Understanding Media*, McGraw-Hill, 1964.

[27] Adriano FABRIS, *Etica per le tecnologie dell'informazione e della comunicazione*, Carocci editore, Roma, 2018.

[28] Ibidem.

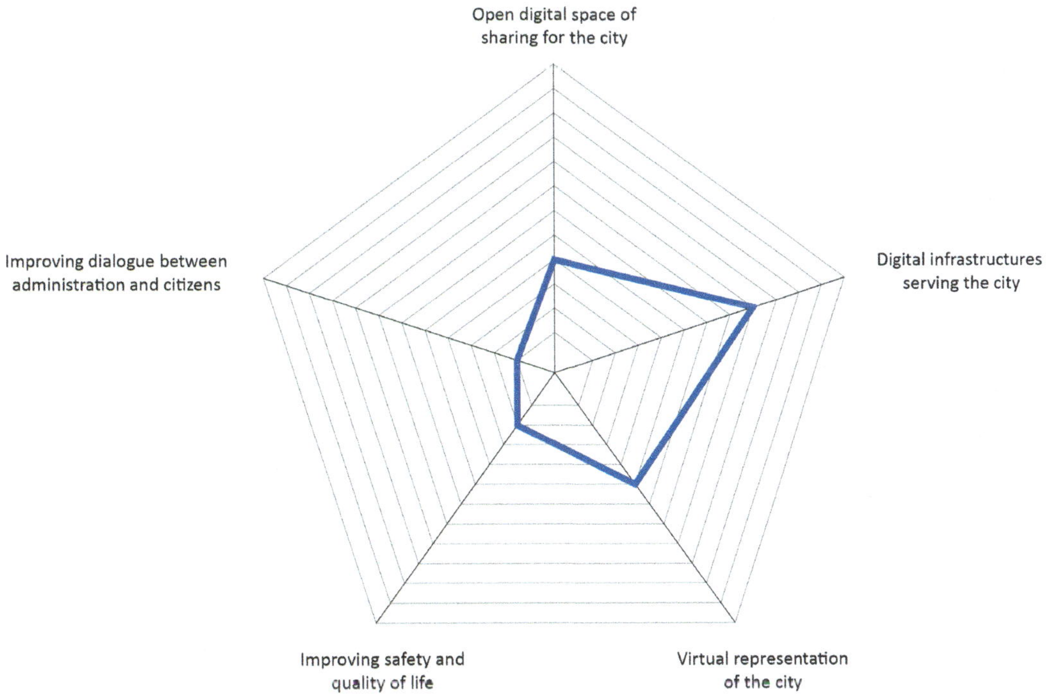

Diagram 1 Key attributes of digital city definitions

3 Digital City

Digital City
Neologism

Definition (Diagram 1)	Source
Urban settlement with technology and telematic connections through which citizens can participate in civic networks and use information and services provided by the administration.	Translated from Italian *Enciclopedia Treccani online*
A digital city is substantively an open, complex and adaptive system based on computer network and urban information resources, which forms a virtual digital space for a city. It creates an information service marketplace and information resource deployment center[29]	Q. Li, S. Lin, 2001

(continued)

Definition (Diagram 1)	Source
Digital cities would then constitute ways in which infrastructure and behavior are being augmented by hardware, software, and data in terms of how these functions are being embodied in smart highways and intelligent buildings as well as in new forms of electronic service delivery[30]	M. Batty, 2001
The concept of digital cities is to build an arena in which people in regional communities can interact and share knowledge, experiences, and mutual interests. Digital cities integrate urban information (both achievable and real-time) and create public spaces in the Internet for people living/visiting in/at the cities[31]	T. Ishida, 2002
A digital city has at least two plausible meanings: (1) a city that is being transformed or re-oriented through digital technology and (2) a digital	D. Schuler, 2002

(continued)

[29] Qi LI, Shaofu LIN, *Research on digital city framework architecture*, IEEE International Conferences on Info-Tech and Info-Net, vol. 1, Proceedings ICII, 2001, pp. 30–36.

[30] Michael BATTY, *Contradictions and Conceptions of the Digital City*, Editorial Environment and Planning B, 28, 2001, pp. 479–480.

[31] Toru ISHIDA, *Digital City Kyoto: Social Information Infrastructure for Everyday Life*, Communications of the ACM 45(7), July 2002, 76–81.

Definition (Diagram 1)	Source
representation or reflection of some aspects of an actual or imagined city[32]	
The digital city is as a comprehensive, web-based representation, or reproduction, of several aspects or functions of a specific real city, open to non-experts. The digital city has several dimensions: social, cultural, political, ideological and also theoretical[33]	H. Couclelis, 2004
U-City, a "ubiquitous city" is defined as a next generation urban space that includes a integrated set of ubiquitous services: a convergent form of both physical and online spaces. These services ultimately aim to enhance quality of life factors, such as convenience, safety and welfare[34]	O. Kwon and J. Kim, 2007
The term 'digital city' denotes an area that combines broadband communication infrastructure with flexible, service-oriented computing systems. These new digital infrastructures seek to ensure better services for citizens, consumers, and businesses in a specific area. The geographical range of digital cities varies from a small part of a city to highly populous metropolises[35]	N. Komninos, 2008
The term Digital City refers to: a connected community that combines broadband communications infrastructure; a flexible, service-oriented computing infrastructure based on open industry standards; and, innovative services to meet the needs of governments and their employees, citizens and businesses. [...] The people involved in the digital/virtual community formed over a digital city platform can interact and share knowledge, experience and mutual interests. The goal of a Digital City is to create an environment for information sharing, collaboration, interoperability	G.S. Yovanof, G.N. Hazapis, 2009
& seamless experience for all its inhabitants anywhere in the city[36]	
Digital City does not refer to a specific urban entity or formal communications mechanism. Instead a functional approach describes four interdependent action types that are highly reliant upon the kinds of digital information outlined above: 1. *supporting data and information*—through, for example advanced analytical interfaces that may emphasize the spatial and temporal nature of information; 2. *providing a communication infrastructure*—physical or virtual means for enabling information flows; 3. *delivering value added information and innovative*—these are likely to synthesize data from a range of sources, be location based and may include analytical interfaces; 4. *using virtual environments in planning, decision-making and analysis*—when data collected by citizens are used in the process of modelling or digitally recorded citizen behavior is influenced by formal planning an analysis a feedback loop is completed. Each of the previously listed action types may contribute here[37]	J. Dykes, 2010
Digital cities, from digital representation of cities, virtual cities, digital metaphor of cities, cities of avatars, second life cities, simulation (sim) city[38]	H. Shaffers et al., 2011
The Digital City is aimed at achieving the goal of improving the quality of life of citizens by improving the dissemination of information, communication and relations in particular between public administration and citizens, i.e. it seems to be a strategy	Translated from Italian R. P. Dameri, L. Giovannacci, 2015

(continued)

[32] Doug SCHULER, *Digital cities and digital citizens*, in M. TANABE, P. VAN DEN BESSELAAR, T. ISHIDA (Eds.), *Digital Cities II: Computational and Sociological Approaches*, Springer Berlin, Heidelberg, 2002, pp. 71–85.

[33] Helen COUCLELIS, *The Construction of the Digital City*, Environment and Planning B: Planning and Design 2004, volume 31, 2004, pp. 5–19.

[34] Ohbyung KWON, Jihoon KIM, *A methodology of identifying ubiquitus smart services for U-city development*, in J. INDULSKA et al. (Eds.), *Ubiquitous Intelligence and Computing, 4th International Conference, UIC 2007, Hong Kong, China, July 11–13 2007, Proceedings*, Springer, Berlin-Heidelberg, 2007, pp. 143–152.

[35] Nicos KOMNINOS, *Intelligent cities and globalization of innovation networks*, Routledge, London, 2008, p. 227.

[36] Gregory S. YOVANOF, George N. HAZAPIS, *An Architectural Framework and Enabling Wireless Technologies for Digital Cities & Intelligent Urban Environments*, Springer, Wireless Personal Communications, 49, 3, 2009, pp. 445–463.

[37] Jason Dykes, *GeoVisualization and the Digital City*, Computers, Environment and Urban Systems, Vol. 34, Elsevier, 2009, pp. 443–451.

[38] Hans SCHAFFERS, Nicos KOMNINOS, Marc PALLOT, Brigitte TROUSSE, Michael NILLSON, Alvaro OLIVEIRA, *Smart Cities and the Future Internet: Towards Cooperation Frameworks for Open Innovation*, in John DOMINGUE et al. (Eds.), *The Future Internet*, Future Internet Assembly 2011: Achievements and technological promises, Springer, 2011, p. 434.

Definition (Diagram 1)	Source
aimed at stimulating cooperation, knowledge sharing, and the pursuit of common interests between the public administration and the inhabitants of the city. [...] The Digital City is the digitised city based on the use of digital technologies by all the actors (public bodies, businesses, citizens) and on active behaviour by citizens with regard to the use of information and digital services, both public and private[39]	
Variants	*Source*
Wired cities refer literally to the laying down of cable and connectivity not itself necessary smart[40]	R.G., Hollands, 2008
Digital environments collecting official and unofficial information from local communities and delivering it to the public via web portals are called information cities[41]	L. Anthopoulos, P. Fitslis, 2010

Definition Proposed by the PhD Students of the Smart City 4.0 Sustainable Lab

The Digital City is the space of interaction within an urban community, understood as a substratum of the real city, where a network of digital individuals can communicate and share knowledge by integrating physical space with virtual information. It is a space within which the socio-technical and physical-virtual dimensions of the city tend to intertwine and overlap thanks to digital infrastructures. The impact of the Digital City on the individual and the community is constant and ubiquitous (ubiquitous computing) and influences habits and patterns of perception of reality in a deep-rooted way. The Digital City can also be understood as a virtual representation or complete reproduction of the real city, based on the web, open to all citizens for the improved use and accessibility of public space. The Digital City model is aimed at improving the quality of life of citizens through free access to information, an increase in the supply of services and the simplification of relations between citizens and public decision-makers. Digitisation is a key factor in enabling individual participation in the connected city and the ultimate aim is to make cities more sustainable in terms of mobility, urban logistics, new technologies for safety and security, incentivising the use of renewable energy sources and implementing systems for resilience to hazard events. However, it is necessary to argue that the digital city cannot establish itself as the replacement of the real city in which there are unique values that cannot be replaced by the technological means

(continued)

[39] Renata Paola DAMERI, Lorenzo GIOVANNACCI, *Smart city e Digital city. Strategie urbane a confronto*, Franco Angeli, Milano, 2015.

[40] Robert G. HOLLANDS, *Will the real Smart City please stand up? Intelligent, progressive or entrepreneurial?*, Analysis of Urban Change, Theory, Action Volume 12, 2008—Issue 3, pp. 303–320.

[41] Leonidas ANTHOPOULOS, Panos FITSLIS, *From digital to ubiquitous cities: defining a common architecture for urban development*, in Intelligent Environments (IE), 2010 SIXTH INTERNATIONAL CONFERENCE ON, 2010, pp. 301–306.

4 Smart City

Smart City
Neologism

Definition (Diagram 2)	Source
A city or town where communication and IT are used to help manage the city and its services.	*Macmillan Dictionary*
An expression currently used for urban planning strategies related to innovation and in particular to the opportunities offered by new communication technologies to improve citizens' quality of life. The 'smart city' concept starts from the assumption of distributed intelligence, inscribed in public space, in a sort of redefinition of the ancient genius loci. The smart city increasingly tends to relate the material infrastructure of cities to the human, intellectual and social capital of those who inhabit them. A capital that is enhanced and systemised by the intangible infrastructures of the web. In the smart city, telematic connectivity is therefore considered a source of growth and urban development to promote the idea of an inclusive city, through the promotion of new forms of social cohesion. A city that is sustainable in terms of ecological control and	Translated from Italian *Enciclopedia Treccani online*

(continued)

Definition (Diagram 2)	Source
energy-saving measures, and aims to optimise mobility and safety solutions through territorial innovation practices. A smart city is such in direct proportion to the quality of the active participation of its citizens, in a context in which the city's public space is interpreted with particular forms of social creativity, such as those of urban experience in the interaction between web and territory	
A city that monitors and integrates conditions of all of its critical infrastructures, including roads, bridges, tunnels, rail/subways, airports, seaports, communications, water, power, even major buildings, can better optimize its resources, plan its preventive maintenance activities, and monitor security aspects while maximizing services to its citizens[42]	R. E. Hall, 2000
Smart city as a city that makes conscious effort to innovatively employ ICTs in support of a more inclusive, diverse and sustainable urban environment[43]	California Institute for Smart Community, 2001
A city where the ICT strengthen the freedom of speech and the accessibility to public information and services[44]	H. Partridge, 2004
A city well performing in a foward-loking way in economy, people, governance, mobility,	R. Giffinger et al., 2007

(continued)

Definition (Diagram 2)	Source
environment and living, built on the smart combination of endowments and activities of self-decisice, independent and aware citizens[45]	
The Smart cities initiative aims to improve energy efficiency and to step up the deployment of renewable energy and climate change policy. This initiative will support cities and regions that take pioneering measures to progress towards a radical reduction of green-house gas emissions through the sustainable use and production of energy. It will bring the cities involved to the forefront of the development of the low-carbon economy[46]	EU, 2007
They are territories with a high capacity for learning and innovation, which is built-in: (1) the creativity of their population; (2) their institutions for knowledge creation; and (3) their digital infrastructure and services for communication and knowledge management. The distinctive characteristic (and ultimate measure) of intelligent cities is their performance in the field of innovation, because innovation and solving of new problems are unique features of intelligence. Intelligent cities and regions constitute advanced territorial systems of innovation, in which the	N. Komninos, 2008

(continued)

42 Robert E. Hall, *The vision of a smart city*, in Proceedings of the 2nd International Life Extension Technology Workshop, Paris, France, September 28, 2000.

43 California Institute for Smart Community, Ten Steps to Becoming a Smart Community, 2001.

44 Helen Partridge, *Developing a human perspective to the digital divide in the smart city*, In Proceedings of the Biennial Conference of Australian Library and information Association, Queensland, Australia, 2004, Sep 21–24.

45 Rudolf Giffinger, Christian Fertner, Hans Kramar, Robert Kalasek, Nataša Pichler-Milanovic, Evert Meijers, *Smart cities. Ranking of European medium-sized cities*, Vienna, Austria, Centre of Regional Science (SRF), Vienna University of Technology, 2007.

46 European Union, *Communication from the commision to the council, the european parlament, the european economic and social committee of the regions. A european strategic energy technology plan (set-plan)–towards a low carbon future*. Brussels. 2007.

Definition (Diagram 2)	Source
institutional mechanisms for knowledge creation and application are facilitated by digital spaces and online tools for communication and knowledge management[47]	
In the IBM vision three 'I's are the hard core of any smart city: instrumented, interconnected and intelligent Instrumented refers to the capability of capturing and integrating live real-world data through the use of sensors, meters, appliances, personal devices, and other similar sensors Interconnected refers to the integration of these data into a computing platform that allows the communication of such information among the various city services Intelligent refers to the inclusion of complex analytics, modeling, optimization, visualization services and artificial intelligence to make better operational decisions[48]	S. Palmisano for IBM, 2008
We believe a city to be smart when investments in human and social capital and traditional (transport) and modern (ICT) communication infrastructure fuel sustainable economic growth and a high quality of life, with a wise management of natural resources, through participatory governance.[49]	A. Caragliu, C. Del Bo, P. Nijkamp, 2009

(continued)

[47] Nicos KOMNINOS, Intelligent Cities and Globalisation of Innovation Networks, London and New York, Routledge, 2008, p. 113.

[48] Sam PALMISANO (CEO IBM), *A Smarter Planet: The Next Leadership Agenda*, in *Smart city: smart story?*, Governance and economy/2017-11-29.

[49] Andrea CARAGLIU, Chiara DEL BO, Peter NIJKAMP, *Smart City in Europe*, 3rd Central European Conference in Regional Science—CERS, Faculty of Economics, Slovakia, 2009, p. 50.

Definition (Diagram 2)	Source
Smarter Cities are urban areas that exploit operational data, such as that arising from traffic congestion, power consumption statistics, and public safety events, to optimize the operation of city services. The foundational concepts are instrumented, interconnected, and intelligent. Instrumented refers to sources of near-real-time real-world data from both physical and virtual sensors. Interconnected means the integration of those data into an enterprise computing platform and the communication of such information among the various city services. Intelligent refers to the inclusion of complex analytics, modeling, optimization, and visualization in the operational business processes to make better operational decisions. This approach enables the adaptation of city services to the behavior of the inhabitants, which permits the optimal use of the available physical infrastructure and resources, for example, in sensing and controlling consumption of energy and water, managing waste processing and transportation systems, and applying optimization to achieve new efficiencies among these resources. Additional roles exist in intelligent interaction between the city and its inhabitants and further contribute to operational efficiency while maintaining or enhancing quality of life[50]	C. Harrison et al., 2010

(continued)

[50] Colin HARRISON, Barbara ECKMAN, Rick HAMILTON, Perry HARTSWICK, Jayant KALAGNANAM, Jurij PARASZKZAK, Peter WILLIAMS, *Foundations for Smarter Cities*, in "IBM Journal of Research and Development", 54(4), 2010.

Definition (Diagram 2)	Source
Smart cities are those that are combining ICT and Web 2.0 technology with other organizational, design and planning efforts to de-materialize and speed up bureaucratic processes and help to identify new, innovative solutions to city management complexity, in order to improve sustainability and "liveability"[51]	D. Toppeta, 2010
The use of Smart Computing technologies to make the critical infrastructure components and services of a city—which include city administration, education, healthcare, public safety, real estate, transportation, and utilities—more intelligent, interconnected, and efficient[52]	D. Washburn, et al., 2010
A smart city connects the past through the present with the future[53]	J. Vojelij, 2010
Smart city is the product of digital city combined with the Internet of Things[54]	K. Su et al., 2011
The concept of smart cities seen from the perspective of technologies and components has some specific properties within the wider cyber, digital, smart, intelligent cities literatures. It focuses on the latest advancements in mobile and pervasive computing, wireless networks, middleware and agent technologies as they become embedded into the physical spaces of cities. The emphasis on smart embedded devices represents a distinctive characteristic of smart cities compared to intelligent cities, which create territorial innovation systems combining knowledge-intensive activities, institutions for cooperation and learning, and web-based applications of collective intelligence[55]	H. Shaffers et al., 2011
Smart City is a city in which it can combine technologies as diverse as water recycling, advanced energy grids and mobile communications in order to reduce environmental impact and to offer its citizens better lives[56]	Setis-Eu, 2012
A city striving to make itself "smarter" (more efficient, sustainable, equitable and livable)[57]	Natural Resources Defense Council, 2012
A territorial context where the planned and wise use of human and natural resources, appropriately managed and integrated through the numerous ICT technologies already available, allows for the creation of an ecosystem capable of making the best use of resources and providing integrated and increasingly intelligent services[58]	Translated from Italian Agenzia per l'Italia Digitale, 2012

(continued)

[51] Donato TOPPETTA, *The Smart City Vision: How Innovation and ICT can build smart, livable and sustainable cities*, The Innovation Knowledge Foundation, Think! Report 005/2010, 2010.

[52] Doug WASHBURN, Usman SINDHU, Stephanie BALAOURAS, Rachel A. DINES, Nicholas M. HAYES, Lauren E. NELSON, *Helping CIOs Understand "Smart City" Initiatives: Defining the Smart City, Its Drivers, and the Role of the CIO*, Cambridge, MA: Forrester Research, Inc, 2010.

[53] Jan Vogelij, *Some thoughts about smart cities*, Soest, NL, 2010.

[54] Kehua SU, Jie LI, Hongbo FU, *Smart city and the applications*, Electronics, Communications and Control (ICECC), International Conference on, IEEE, 2011.

[55] Hans SCHAFFERS, Nicos KOMNINOS, Marc PALLOT, Brigitte TROUSSE, Michael NILLSON, Alvaro OLIVEIRA, *Smart Cities and the Future Internet: Towards Cooperation Frameworks for Open Innovation*, in John DOMINGUE et al. (Eds.), *The Future Internet*, Future Internet Assembly 2011: Achievements and technological promises, Springer, 2011.

[56] SETIS-EU, 2012, setis.ec.europa.eu/implementation/technology-roadmap/.

[57] Natural Resources Defense Council, 2012

[58] Agenzia per l'Italia Digitale, 2012.

Definition (Diagram 2)	Source
A smart city is a well-defined geographical area, in which high technologies such as ICT, logistic, energy production, and so on, cooperate to create benefits for citizens in terms of well-being, inclusion and participation, environmental quality, intelligent development; it is governed by a well-defined pool of subjects, able to state the rules and policy for the city government and development[59]	R.P. Dameri, A. Cocchia, 2013
Smart city as a city that addresses public problems through the use of ICT, based on a multi-stakeholder partnership[60]	Translated from Italian Committee on Industry, Research and Energy (ITRE), 2014
An immaterial space of infinite horizontality [...] The risk is to observe the city only as a place for sharing services and to lose its connotation as a place for sharing spaces[61]	Translated from Italian A. Zamboni, 2014
A city can be smart if investments in human and social capital, traditional communication infrastructure (transport) and modern ICT are able to support economic growth and improve the quality of life, through wise management of natural resources and through participatory governance[62]	Translated from Italian R. P. Dameri, L. Giovannacci, 2015

(continued)

Definition (Diagram 2)	Source
The performance of a Smart City therefore depends not only on the endowment of physical infrastructure, but also on the availability and quality of communication and sharing of knowledge and social infrastructure. In other words, the smart city is such if it becomes a system, both in the functional aspects of the urban context and in the promotion of inclusive initiatives, in which the web and new technologies become an extension of the human, intellectual and cultural capital of citizens. cultural capital of citizens[63]	Translated from Italian S. Siniscalchi, 2017
A city can be said to be smart if it is technological and interconnected, clean, attractive, reassuring, efficient, open, collaborative, creative, digital and green. […] In the smart city, an ecosystem of sensors collects information from the urban space, on which a series of actuators can then act, modifying it. Data-driven processes transform the city into a test bed; a mixed digital and physical space, unified by a distributed computing platform[64]	Translated from Italian C. Ratti, 2017
The idea of a smart city is that of an interconnected city where everything converges, where people are entirely connected. The future of humanity on earth is represented by this electrical interaction that now goes through a digital phase, but tomorrow will feature the quantum phase	Translated from Italian D. De Kerckhove, 2020

(continued)

59 Renata Paola DAMERI, Annalisa COCCHIA, *Smart City and Digital City: Twenty Years of Terminology Evolution*, 2013.

60 Francesca BATTISTONI, *La mappa delle Smart City Eu e la sinergia con EUROPA2020*, 10 february 2014, https://www.forumpa.it/citta-territori/la-mappa-delle-smart-city-eu-e-la-sinergia-con-europa2020/.

61 Andrea ZAMBONI, *La città intelligente, l'uomo è rinsavito?*, «Domus», n°985 Novembre 2014.

62 Renata Paola DAMERI, Lorenzo GIOVANNACCI, *Smart city e Digital city. Strategie urbane a confronto*, Franco Angeli, Milano, 2015.

63 Silvia SINISCALCHI, *Smart city e governance del territorio. Le potenzialità degli opendata cartografici attraverso alcuni casi di studio*, Bollettino della Associazione Italiana di Cartografia 2017 (160), 69–79, EUT EDIZIONI UNIVERSITÀ DI TRIESTE, 2017.

64 Carlo RATTI, *La città di domani, Come le reti stanno cambiando il futuro urbano*, Einaudi, Torino, 2017.

Definition (Diagram 2)	Source
correlating all the IoT (Internet of Things, made up of sensors everywhere), the IoB (Internet of Behaviors, connecting every move made by everybody), the IoP (Internet of People, bringing potentially everyone in contact with everyone else). This means that nothing can escape this bond of connectedness. This electronic straightjacket condition obviously creates difficulties on the level of individual identities and the autonomy of the person. That has to be negotiated in a culture such as that of the West where individuality is prized above all else (in education, sport, business and employment) and privacy, albeit in sharp decline, is still treasured. People have to recognize that digital interaction brings them to total interconnection, not only in the smart city, but even more so in the metaverse where eye movements need to be recorded in order to be functional. Can we afford to lose our civil autonomy and freedom of conscience that we gained with such strife, torture and bloodshed during the Renaissance? Today mental, physical and social independence is thrusted inside personal portable devices. The alternative is to change the ground, to accept connectedness while maintaining personal values. Ethics lies in figuring out how to harmonize the individual's presence in the world, and how to be useful in the city[65]	

(continued)

[65] Derrick DE KERCKHOVE, *Smart City o Wise Town? Conversazioni sulla città della Quarta Rivoluzione Industriale*, SMART CITY 4.0 Sustainable LAB, Parma, Videointerview with Derrick De Kerckhove, 2020.

Definition (Diagram 2)	Source
The Smart City is not just an urban centre that makes massive use of technology without addressing the issue of how to manage it so as not to create inequalities and ensure rights for all by limiting the digital divide. For me, the Smart City integrates two aspects: the one related to the relevant presence of technology and the one that sees an equal and usable distribution of these resources by all. I believe precisely that an extremely technological city where no one had the possibility of accessing these resources would be no different from a city without technology. The advantage can be seen from the moment everyone starts to access and use technology. Hence the need for a plan for its deployment, i.e. a default-deployment managed by a figure above the parties who can steer the exploitation of technology in the right direction. This is to ensure a fair distribution of technology without it benefiting only those who know it and those who can master it[66]	Translated from Italian A. Broggi, 2020
The dream of the smart city, which through technological solutions finds well-being and finds happiness, I fear is a disappointment. Also because a smart city that stops at this is a smart city in which the problems of the just city, the problems of cohesion, the problems of coexistence, the problems of sharing find limited solutions. So, if I had to give it a journalistic subtitle, I would say that	Translated from Italian G. M. Flick, 2020

(continued)

[66] Alberto BROGGI, *Smart City o Wise Town? Conversazioni sulla città della Quarta Rivoluzione Industriale*, SMART CITY 4.0 Sustainable LAB, Parma, Videointerview with Alberto Broggi, 2020.

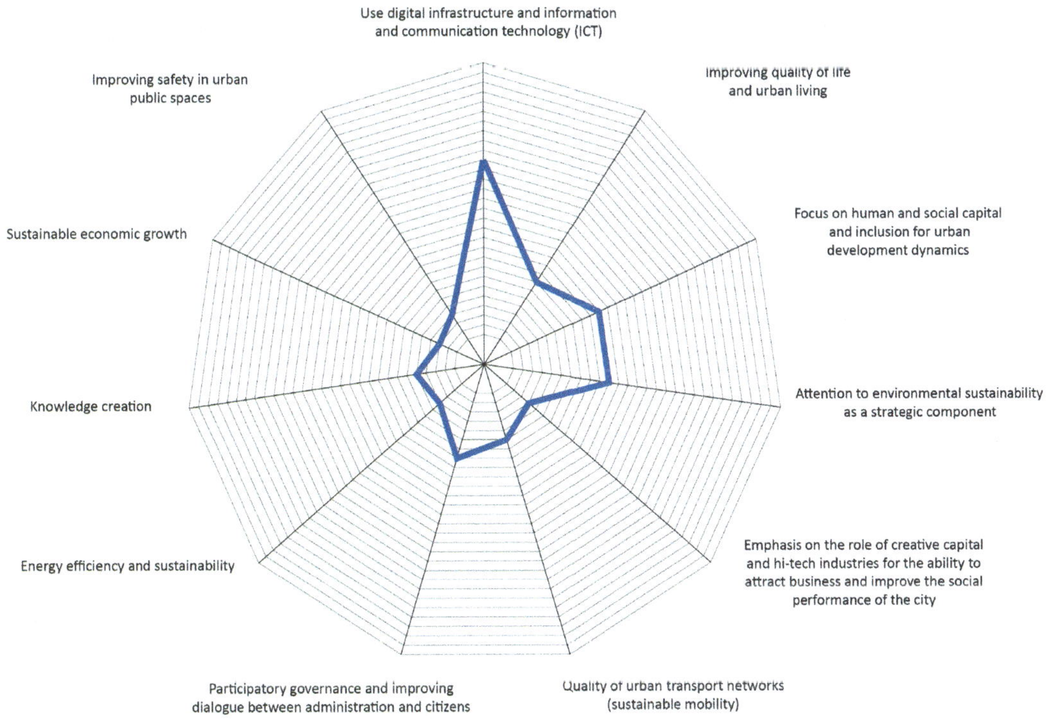

Diagram 2 Key attributes of smart city definitions

Definition (Diagram 2)	Source
the smart city is the city of illusion[67]	
A smart city is a place where traditional networks and services are made more efficient with the use of digital and telecommunication technologies for the benefit of its inhabitants and business. A smart city goes beyond the use of information and communication technologies (ICT) for better resource use and less emissions. It means smarter urban transport networks, upgraded water supply and waste disposal facilities and more efficient ways to light and heat	European Commission

(continued)

Definition (Diagram 2)	Source
buildings. It also means a more interactive and responsive city administration, safer public spaces and meeting the needs of an ageing population[68]	
Definition Proposed by the PhD Students of the Smart City 4.0 Sustainable Lab	
The Smart City concept is circumscribed to a specific geographical area or portion of urbanised territory where actions or strategies have been undertaken to improve the physical and digital environment, for an increase in competitiveness, attractiveness of services and inclusiveness, focusing mainly on six strategic axes: economy, mobility, environment, people, quality of life and governance The application of the Smart City concept envisages a holistic strategy for the development and management of the city through a bottom-up model where individual actions merge into an overall vision. The Smart City is	

(continued)

[67] Giovanni Maria FLICK, *Smart City o Wise Town? Conversazioni sulla città della Quarta Rivoluzione Industriale*, SMART CITY 4.0 Sustainable LAB, Parma, Videointerview with Giovanni Maria Flick, 2020.

[68] https://ec.europa.eu/info/eu-regional-and-urban-development/topics/cities-and-urban-development/city-initiatives/smart-cities_en.

distinguished, in general, by the massive use of information and communication technologies (ICT) that are implemented to make infrastructurally critical city components and services (administration, education, healthcare, public safety, transport, public services) more intelligent, interconnected, efficient, sustainable and accessible
The most impactful innovation actions are applied in the areas of technology, energy and infrastructures (transport and ICT) and are combined with decision-making processes that include social inclusion and the involvement of the population through participatory policies.
Despite the fact that the ultimate goal is to improve the quality of life of citizens, the Smart City concept lacks references to the preservation of the spatial qualities of the city and due to the nature of the quantitative approach, a local development strategy, based on qualitative investigations, cannot be identified with the term Smart City, capable of adequately enhancing the identity of contexts

5 Green City

Green City
Neologism

Definition (Diagram 3)	Source
[...] Eco-cities, or sustainable communities, represent a goal, a direction for community development. [...] The "eco-cities" theme does not stand alone but is situated in a complex array of relevant variations [...] (sustainable development, community economic development, sustainable urban development, appropriate technology, sustainable communities, sustainable cities, bioregionalism, green movement, green cities/communities)[69]	M. Roseland, 1997
Green cities have clean air and water and pleasant street and parks. Green cities are resilient in the face of natural disasters, and the risk of major infectious disease outbreaks in such cities is low. Green cities also encourage green behavior, such as the use of public transit, and their ecological impact is relatively small[70]	M. E. Kahn, 2006
Green City follows the Green Growth which is a new paradigm that promotes economic development while reducing greenhouse gas emissions and pollution, minimizing waste and inefficient use of natural resources and maintaining biodiversity[71]	OECD, 2010
Sustainable city uses technology to reduce CO_2 emissions, to produce efficient energy, to improve the buildings efficiency. Its main aim is to become a green city[72]	L. Batagan, 2011
Green cities are defined as those that are environmentally friendly. [...] The greening of cities requires some, or preferably all, of the following: (1) controlling diseases and their health burden; (2) reducing chemical and physical hazards; (3) developing high quality urban environments for all; (4) minimising transfers of environmental costs to areas outside the city; and (5) ensuring progress towards sustainable consumption[73]	UNEP, 2011
For us the green city is the model of the future, creating urban structures with environments with life-quality. The sustainable green development of cities is a task to be continuously developed, which calls for integrated and regionally coordinated activities of all disciplines[74]	ELCA, 2011
The green city aims at limiting environmental impact and safeguarding natural resources—air, water, soil, energy sources—by	R. P. Dameri, 2015

(continued)

[69] Mark ROSELAND, *Dimensions of the eco-city*, Cities, 14 (4), 1997, pp. 197–202.

[70] Matthew E. KAHN, *Green cities: urban growth and the environment*, Washington, DC, Brookings Institution Press, 2006.

[71] OECD, *Cities and Green Growth*, Issues Paper for the 3rd Annual Meeting of the OECD Urban Roundtable Of Mayors And Ministers, OECD, Paris, 2010.

[72] Lorena BATAGAN, *Smart cities and sustainability models*, Revista de Informatica Economica, 15(3), 2011, pp. 80–87.

[73] United Nation Environment Programme (UNEP), *Towards a green economy: Pathways to sustainable development and poverty eradication. A synthesis for policy makers*, 2011.

[74] European Landscape Contractors Association (ELCA), *Green City Europe—for a better life in European cities*, ELCA Research Workshop, 2011.

Definition (Diagram 3)	Source
employing not only the most advanced technologies, but also the creation of a green culture that promotes virtuous behaviour by citizens[75]	
The concept of "Green City" or "Green Development" is not new. Previously couched within the term "sustainable development," it seeks to integrate environmental, social, and economic considerations within development processes. A Green City or Green Development is an extension of this concept but is understood within the frame of a city's actions and how these actions contribute to a city or urban area advancing as green and sustainable. Green Development considers how to improve and manage the overall quality and health of water, air, and land in urban spaces; its correlation with hinterlands and wider systems; and the resultant benefits derived by both the environment and residents[76]	ADB, 2015
A "Green City" is a city that takes responsible political and societal action in order to achieve high environmental quality, which by itself contributes to human well-being[77]	R. Pace et al., 2016
A city that promotes energy efficiency and renewable energy in all its activities, extensively promotes green solutions, applies land compactness with mixed land use and social mix practices in its planning systems, and anchors its local development in the principles of green growth and equity[78]	O Brilhante, J. Klaas, 2018

(continued)

[75] Renata Paola DAMERI, *Urban Tableau de Bord: Measuring Smart City Performance*, Lecture Notes in Information Systems and Organization, in Lapo MOLA, Ferdinando PENNAROLA, Stefano ZA (ed.), *From Information to Smart Society*, edition 127, pages 173–180, Springer, 2015.

[76] Asian Development Bank (ADB), *Green City Development Tool Kit*, Asian Development Bank, Manila, 2015.

[77] Rocco PACE, Galina CHURKINA, Manuel RIVERA, *How green is a "Green City"? A review of existing indicators and approaches*, Institute for Advanced Sustainability Studies (IASS) Potsdam, December 2016.

[78] Ogenis BRILHANTE, Jannes KLAAS, *Green City Concept and a Method to Measure Green City Performance over Time Applied to Fifty Cities Globally: Influence of GDP, Population Size and Energy Efficiency*, MDPI Journals, Sustainability Volume 10, (6), 2018.

Definition (Diagram 3)	Source
[...] There is a third type of city that is very important, whose connecting link is the reconciliation between man and the environment. On the one hand between man and nature and on the other hand the recognition of the fundamental principle of solidarity. Remember that there can be no equality without solidarity, and solidarity is, quoting the Constitution, that which makes it possible to achieve a balance of equal social dignity through the removal of obstacles that result in abuse or discrimination I am admired by the effort with which Pope Francis reproposed the Canticle of the Creatures reminding us of the values we had forgotten about the importance of nature. We celebrated the 2015 World Expo by taking as our emblem not a thousand-year-old tree, of which we are rich in Italy, but a nickel, steel and LED tree. I would have preferred a thousand-year-old olive tree that we have in Puglia to represent the Expo. The third hypothesis is therefore the Green City, which we have tried to do with a few small experiments. Urban green is not just the tree house or the apartment block with trees in every balcony. Urban green is the royal gardens, which then become everyone's heritage as ecological parks. A beginning of rethinking the enjoyment of space. There is a great mission for architects, which is to graduate a capacity for distinction between public and private space, between everyone's enjoyment of space and the space in which everyone withdraws. It is precisely the discourse of the enhancement of urban greenery that is very interesting[79]	Translated from Italian G. M. Flick, 2020
Urban regeneration according to the green city approach takes ecological quality as a strategic priority in order to ensure sustainability and resilience of programmes and projects in the era of climate crisis, soil scarcity and other natural resources. Urban regeneration in a green city key challenges both its reductive version based on punctual interventions,	Translated from Italian Green City Network

(continued)

[79] Giovanni Maria FLICK, *Smart City o Wise Town? Conversazioni sulla città della Quarta Rivoluzione Industriale*, SMART CITY 4.0 Sustainable LAB, Parma, Videointerview with Giovanni Maria Flick, 2020.

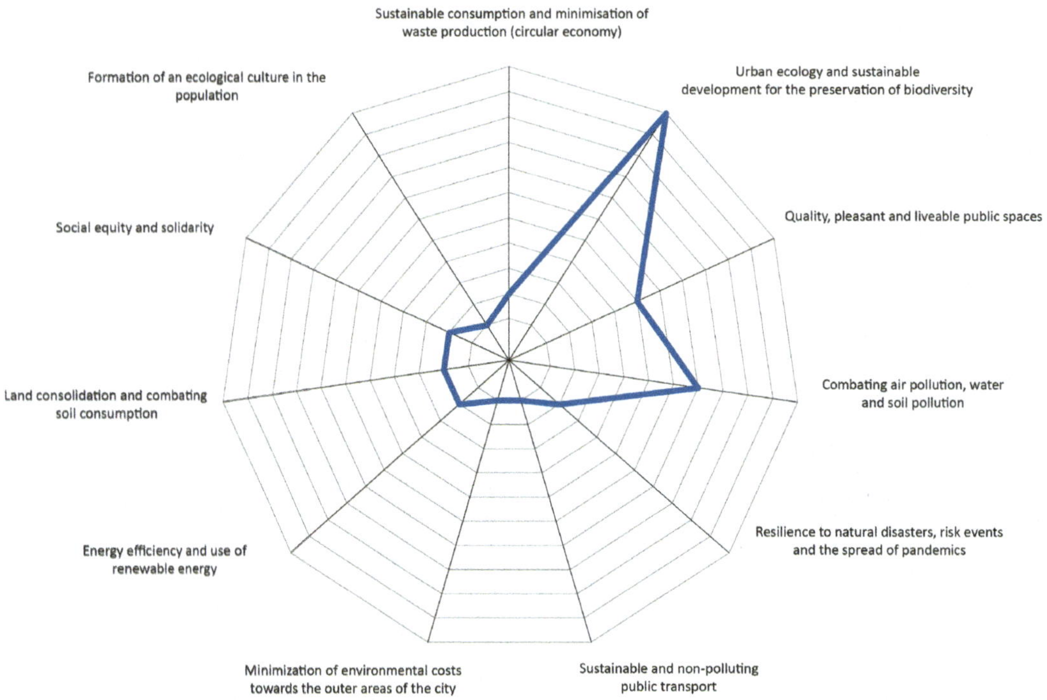

Diagram 3 Key attributes of green city definitions

Definition (Diagram 3)	Source
lacking framework, vision and the necessary ecological quality—and its general-general version—based yes on very broad contents (economic, social, cultural, settlement, infrastructural, etc.), but lacking priority and coherence with the ecological challenges that cannot be postponed and, for this reason, weak and qualitatively inadequate. Urban regeneration, according to the green city model, aims instead to make the different and connected aspects of high ecological quality the actual priorities on which, with an integrated and multi-sectoral approach, to base programmes and intervention projects[80]	

(continued)

[80] Green City Network, *Carta per la rigenerazione urbana delle green city*, https://www.greencitynetwork.it/documenti/.

Definition Proposed by the PhD Students of the Smart City 4.0 Sustainable Lab
The Green City is a city that recognises itself in the green, accessible and welcoming dimension of public space thanks to an integrated system of routes for soft and sustainable mobility (greenways) that connect the different parts of the city in a pleasant and safe manner. Although the term 'Green' denotes a focus on the environmental aspects of air, water and soil quality and energy efficiency, established examples recognised by international rankings recognise in green cities virtuous policies of social inclusion, safety, sustainable mobility and education on healthy lifestyles. In the Green city, urban planning is guided by the principle of the sustainable development of the city where the quality of public space assumes a role of primary importance. The compactness of the urban territory, the fight against unjustified expansion and land consumption are the key principles of strategic choices for development and growth. The Green city is also designed to be resilient to natural disasters and the development of epidemics by defining the strategy of minimal environmental impact through the education and commitment of the inhabitants to minimise both the required inputs of energy, water and food and the wasteful output of heat, waste, pollution of water, soil, air and climate-changing emissions

6 Wise City

Wise City
Neologism

Definition (Diagram 4)	Source
The city that does not eliminate human history, that recovers the valid values of the past and places them in a new relationship with the possible values of the future, and that guarantees the multiplicity of choices to all citizens[81]	C. Aymonino, 1965
That future thinking about cities should focus not on developing smart cities but on creating wise cities. How do we do this? There are many ways, but one possibility is to tap into the resources of local universities. In many cities, universities are the sleeping giants of place-based leadership and social innovation. However, the giant is waking up. Across the world higher education is undergoing significant change and, as part of the rethinking of the role of universities in modern society, the very nature of scholarship is being reconsidered. It is encouraging to note that a growing number of scholars in a wide range of disciplines now see active engagement with the city as a splendid way to advance knowledge and understanding, invent new theories as well as contribute to public purpose. Universities can, perhaps, assist in helping public policy makers and activists deepen understanding of public learning and radical innovation in the modern city[82]	R. Hambleton, 2015
Be wise, not smart: the term 'smart cities' is abused. What's the alternative—a dumb city? Advocates are often linked to a salesman at a Silicon Valley tech company. City halls have always collected information (or 'big data') and tried to glean truths from it. Just because techies can now crunch the data quicker doesn't make it a change that should invoke evangelical tones. The wisest cities also focus on 'small data'—the incidentals that actually make our lives better. [...] It is not always about big ideas, the tech solution or the grand plan. Sometimes all you need to make a better city better is some humanity, a sense of scale and keen citizens[83]	A. Tuck, 2018
Also called Living Smart City, it is a city that supports the creation of self-sufficient communities with services (grocery shops, parks, cafes, sports facilities, health centres, schools, coworking) nearby, reachable by bicycle, to reduce pollution and stress. Reinventing the idea of urban proximity, the aim is to offer what communities need close to their doorstep to ensure an 'ecological transformation' of cities in the post-vehicle era[84]	A. Hidalgo, 2020
[...] Wisdom is not the same as intelligence and the two terms do not always coincide as the Greeks taught us when they used the word phronesis. Talking about a wise city is better than talking about a smart city. To clarify the point, it is good to remember Cicero who taught us about the distinction between civitas and urbs some two thousand years ago. The *civitas* is the city of souls and the *urbs* is the city of stones. The *urbs* model sees the city as a space, the *civitas* model sees the city as a place. While space is a geographical category that is measured by quantitative variables, place, on the other hand, refers to a living environment. This is why it is important to speak of a wise city because it means recovering that ancient distinction that the Romans had already provided us with. The problem is that if we turn to history, as architects know very well, we have transformed the model of the *civitas* into the model of the *urbs*, and now we are paying the consequences. Recovering these roots that are typical of our cultural matrix is, at this historical moment, a work of great relevance. We need to return to thinking about the city as a place of organised coexistence. The problem is at the heart of Book XIX of the City of God, where Augustine takes up the	Translated from Italian S. Zamagni, 2020

(continued)

[81] Carlo AYMONINO, *Origini e sviluppo della città moderna*, Marsilio, Padova, 1965.

[82] Robin HAMBLETON, *From smart cities to wise cities*, Paper for the AAG Annual Meeting, Chicago, Illinois, 21–26, April 2015.

[83] Andrew TUCK (Eds.), *The Monocle Guide to Building Better Cities*, Gestalten, Berlin 2018.

[84] Anne HIDALGO, *La Ville Du Quart d'Heure*, Paris, 2020.

Definition (Diagram 4)	Source
comparison with Cicero. What the bishop of Hippo makes abundantly clear is that the city is a public place inhabited by a people. But what makes it possible for different people to live together, transforming a shapeless aggregate of individuals into a true people? Compared to the Ciceronian answer evoking the primacy of justice, Augustine identified concord as the glue that gives birth to a people: love (from the Latin *cumcorde*, to put hearts together) not only does not weaken institutions or destabilise the political sphere, but rather represents its true generative principle[85]	
I like the idea of the Wise Town. It is an imaginable concept that could really come to fruition as well as that of the Proud Town. People need to recover the pride of their city. The first lockdown caused by the Covid-19 outbreak was a lesson in wisdom. We glimpsed the benefits that can emerge from a wise town. Eliminated for a while were the excesses of traffic, noise and consumption but as soon as the forced confinement ended everyone forgot all about it[86]	Translated from Italian D. De Kerckhove, 2020
[...] Then there is the stage of the wise city: the Wise city, that is, a city that achieves balance. Beware, however, because there is a very great danger: the city of balance, which requires rules to guarantee it, could only end up becoming the city of institutions, of bureaucracy, of the relationship between local autonomies and central power. I believe that the wise city is still a utopia and we should work hard to get there[87]	Translated from Italian G. M. Flick, 2020
What we call smart is the development and colonisation of our milieux by a series of communication and transmission technologies. It's a question of connecting the distinction	Translated from Italian S. Marot, 2020

(continued)

[85] Stefano ZAMAGNI, *Smart City o Wise Town? Conversazioni sulla città della Quarta Rivoluzione Industriale*, SMART CITY 4.0 Sustainable LAB, Parma, Videointerview with Stefano Zamagni, 2020.

[86] Derrick DE KERCKHOVE, *Smart City o Wise Town? Conversazioni sulla città della Quarta Rivoluzione Industriale*, SMART CITY 4.0 Sustainable LAB, Parma, Videointerview with Derrick De Kerckhove, 2020.

[87] Giovanni Maria FLICK, *Smart City o Wise Town? Conversazioni sulla città della Quarta Rivoluzione Industriale*, SMART CITY 4.0 Sustainable LAB, Parma, Videointerview with Giovanni Maria Flick, 2020.

Definition (Diagram 4)	Source
that you make between smart and wise: as you well know, smart means clever, intelligent, fast. It basically evokes technical solutions that are scalable, that work really well in any situation. They are technical routines that have a certain level of effectiveness in almost all situations, so they are often unwieldy. The word wise, means a form of intelligence that also involves prudence, equilibrium, a balance of things. Wise is always a consideration that is fine and careful, it refers to the area of attention, the situation that we have to face. There's something very casuistic about wisdom. What worries me is the danger that we'll rush into scalable solutions that can be bought, and that can work together with a fairly rapid adaptation. And therefore forget, aside from the effectiveness of the instruments that allow it to work, the very reason for the milieux that are transformed by these solutions[88]	
Individual journeys can be significantly reduced if we succeed in combining a series of related factors. We will be able to create a system of protected areas where we can move about on foot and by bicycle every day if we can choose to cross a rearranged city that demonstrates a series of related capacities: if the individual businesses on the street make maximum use of the public spaces they overlook, if self-driving vehicles reduce individual multiplication of today's cars by enabling economic and ecological sharing, if centralised logistics reduce journeys to shops, if the willingness of traders to provide home deliveries restores personal relationships and encourages people to return to the shops, and if remote technology removes the need to attend many appointments in person. If we make the most of all these life conditions, we will combine a return to the habits of a seemingly forgotten past with the intelligent use of technologies that we thought were a thing of the future. We can then imagine a life where we can pick and choose the experiences that interest us. This emergency will have	D. Costi, 2021

(continued)

[88] Sebastien MAROT, *Smart City o Wise Town? Conversazioni sulla città della Quarta Rivoluzione Industriale*, SMART CITY 4.0 Sustainable LAB, Parma, Videointerview with Sebastien Marot, 2020.

Definition (Diagram 4)	Source
allowed us to rediscover the beauty of living in cities, also through the convenience of delegating tasks that can be performed remotely. Having become aware of the failings that we have suffered and the opportunities that we have enjoyed, this scenario is today considerably closer than yesterday. We can then imagine a situation where the old habits of traditional European city dwellers are enhanced with the new opportunities provided by the permanent interconnection and mandatory digitalisation that we have experienced. The smart city can therefore also become a wise town. It can enjoy the best conditions of digital accessibility and immediate connections with the world, even in the most isolated districts. At the same time, it can enjoy the particularities and services of a small town, even in the more extensive and undifferentiated contexts[89]	
Variants	*Source*
Intelligent cities are territories with high capability for learning and innovation, which is built-in the creativity of their population, their institutions of knowledge creation, and their digital infrastructure for communication and knowledge management[90]	N. Komninos, 2006
The idea is to hack the city: to open up traditionally closed computer systems and break the entrenched mentality of optimising urban spaces. In short, you can empower people to take an active role in their environment. You can use open-source technologies to aggregate knowledge, skills, and ideas from a wide and heterogeneous citizenry and make the changes tangible. [...] Smart cities should be rethought as ‘senseable cities’, a definition that emphasises the centrality of humans within them[91]	Translated from Italian C. Ratti, 2017

(continued)

Definition (Diagram 4)	Source
A Knowledge City is a city that aims at a knowledge-based development, by encouraging the continuos creation, sharing, evaluation, renewal and update of knowledge. This can be achieved through the continuos interaction between them and other cities’ citizens. The citizens’ knowledge-sharing culture as well as the city’s appropriate design, IT networks and infrastructures support these interactions[92]	K. Ergazakis et al., 2006
The Smart City is not about spirituality, which is why I prefer to say Care City. In the Care City, the focus is not only on efficiency and productivity. The city must go back to the Renaissance model, to the time when the first hospitals, hospices, and confraternities were born [...] We must be careful about using words, and Smart City is a term that must be abandoned because it is wrong, unless we want to agree with the transhumanists[93]	Translated from Italian S. Zamagni, 2020

Definition Proposed by the PhD Students of the Smart City 4.0 Sustainable Lab
The wise city is the evolution of the smart city concept that brings together the components of sustainability and knowledge-based development. In the wise city, technology is above all a support for the citizen who becomes the centre and protagonist of choices and not just the recipient of technological development In the wise city, a wise use of resources is made, while strengthening local participation and welfare through co-designing the urban space with citizens according to their needs. This participatory process leads to the reduction of social exclusion and loneliness, making cities more suitable for children, the elderly and the disabled. In fact, the wise city focuses on quality of life, where no one feels excluded thanks to participatory processes that also increase trust in institutions. The wise city seeks to fuse informal knowledge derived from

(continued)

[89] Dario COSTI, *Designing the City of People 4.0. Reflections on strategic and sustainable urban design after Covid-19 pandemic*, Series The City Project. Strategies for Smart and Wise Sustainable Urban Design 1, Springer, Berlin, 2021.

[90] Nicos KOMNINOS, *The Architecture of Intelligent Cities*, Intelligent Environments 06, Institution of Engineering and Technology, pp. 13–20, 2006.

[91] Carlo RATTI, *La città di domani, Come le reti stanno cambiando il futuro urbano*, Einaudi, Torino, 2017.

[92] Kostas ERGAZAKIS, Kostas METAXIOTIS, John PSARRAS, *Knowledge cities: The answer to the needs of knowledge-based development*, VINE: The journal of information and knowledge management systems, Vol. 36 No. 1, 2006, pp. 67–84, 2006.

[93] Stefano ZAMAGNI, *Smart City o Wise Town? Conversazioni sulla città della Quarta Rivoluzione Industriale*, SMART CITY 4.0 Sustainable LAB, Parma, Videointerview with Stefano Zamagni, 2020.

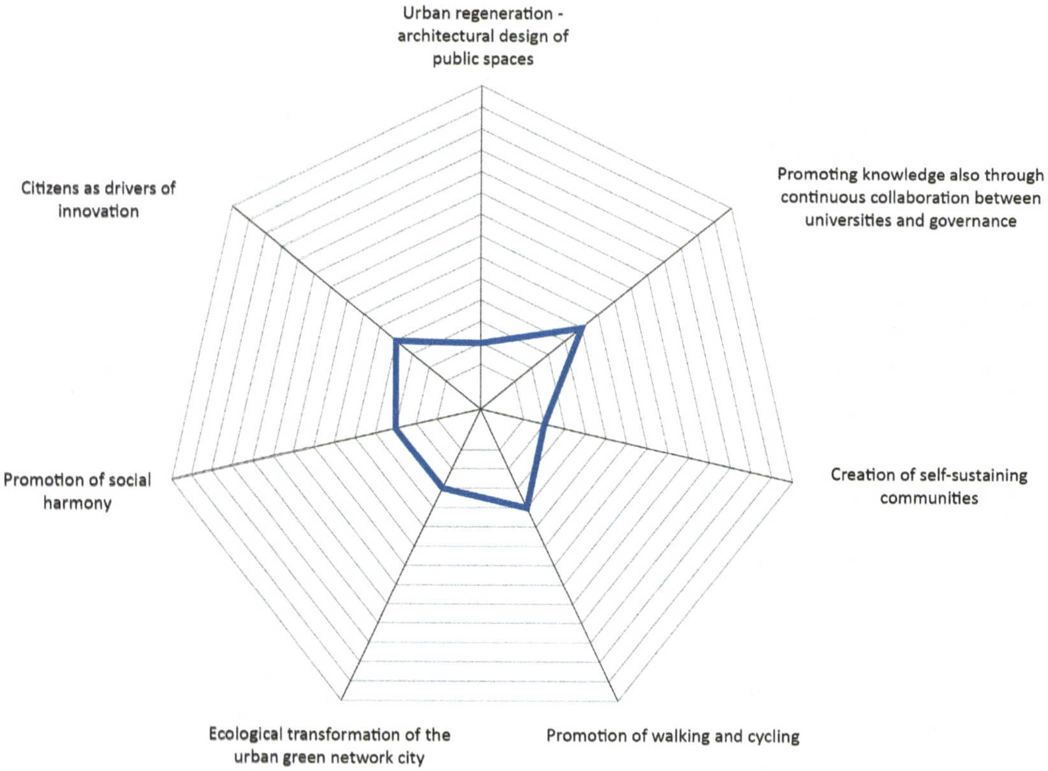

Diagram 4 Key attributes of wise city definitions

the emotional, identity, personal and social aspects of the city with formal, scientific and professional knowledge, also derived from the use and study and data as new technologies allow. The term places the idea of critical knowledge and the expression of judgement on processes at the centre, assuming an idea of city development to which only then can the tools traditionally pertaining to the smart city concept be put. In the wise city, knowledge is promoted through collaboration between local government and universities, encouraging continuous creation, sharing, renewal and updating through territorial innovation systems that combine knowledge-intensive activities and institutions. In the wise city, an idea of the city of people is promoted, green, accessible and welcoming through an urban planning process that focuses on the quality of public space, the ecological transformation of settlements and the pleasantness of walking or cycling as an opportunity for social experience, towards a paradigm shift in individual and collective urban mobility

7 The City of People 4.0

The City of People 4.0
Neologism

Definition	Source
I thought it might be useful to start by outlining the basic issues to be tackled and with a synthetic and strategic reflection on the contemporary city driven by the health emergency. "The City Project" series of books that we have been planning since last year aims, in fact, to ignite a debate on applied research in the international scientific context and on operational developments that promote urban regeneration and the sustainable transformation of contemporary cities in the era of the Fourth Industrial Revolution, centred on the people who inhabit them. The series is	D. Costi, 2021

(continued)

Definition	Source
inspired by the idea of the City of Man that was developed by the enlightened entrepreneur Adriano Olivetti and pursued by the architect Ernesto Nathan Rogers in the post-war period. By refining, clarifying and in some ways surpassing the still rather uncertain and multiform concept of smart city that we have inherited, this editorial effort seeks to present the many strategies and contributions that humanistic culture and scientific culture can bring together in realising the City 4.0—the City of People in the fourth industrial revolution—in line with the objectives of the United Nations Agenda 2030 for Sustainable Development […] We propose that the architectural design of public spaces be the tool to implement a smart and wise city, reorganising the public city around people, organising urban parks and vacant land, and taking advantage of the technological innovations of the fourth industrial revolution that will make it possible to create a city that is smart but also wise, and of big data as well as critical reflection: the City of Man 4.0. If the response is adequate, it will perhaps be possible to resolve one final conflict—the one between government and community, between policies and projects, between pretending and doing. The resurgence of an emulative dialectic of recovery between public and private sectors is the engine that can transform the primarily cultural fabric of society and steer a transformation of the physical structure of its cities, with efficient integrated tools and according to new values that rediscover ancient principles. Now is the time[94]	

[94] Dario COSTI, *Designing the City of People 4.0. Reflections on strategic and sustainable urban design after Covid-19 pandemic*, Series The City Project. Strategies for Smart and Wise Sustainable Urban Design 1, Springer, Berlin, 2021.

Federico Diodato is a PhD in architecture and design cultures from the University of Bologna in co-operation with the Université Paris Est Créteil - Val de Marne (UPEC).

Andrea Fanfoni is a PhD in architectural and urban design from the Department of Engineering and Architecture at the University of Parma.

Emanuele Ortolan is a PhD in architectural and urban design from the Department of Engineering and Architecture at the University of Parma.

Smart City or Wise Town? Conversations on the City of the Fourth Industrial Revolution

Dario Costi, Federico Diodato, Andrea Fanfoni, and Emanuele Ortolan

Abstract

Smart City 4.0 Sustainable LAB identified a number of internationally recognised experts who have studied and are studying city issues from very different perspectives. They were asked various questions, some of them common, that question them on the role of technology, the meaning of the Smart City and the possibility that beyond the fast city promised by the Smart City, there could be a Wise Town, i.e. a slow and reflective city, attentive to social relations.

Contributions come from civil economist Stefano Zamagni, media sociologist Derrick De Kerckhove, autonomous driving expert Alberto Broggi, constitutional jurist Giovanni Maria Flick and philosopher Sebastién Marot, who have accepted this invitation and provided very valuable insights for the research that the laboratory is conducting on the critical-cultural foundations of the contemporary city.

D. Costi · A. Fanfoni · E. Ortolan (✉)
University of Parma, Parma, Italy
e-mail: emanuele.ortolan@unipr.it

D. Costi
e-mail: dario.costi@unipr.it

A. Fanfoni
e-mail: andrea.fanfoni@unipr.it

F. Diodato
Alma Mater Studiorum University of Bologna, Bologna, Italy
e-mail: federico.diodato2@unibo.it

Conversation with the Economist Stefano Zamagni

Introduced and interviewed by Dario Costi
27 October 2020

DARIO COSTI Stefano Zamagni is an Italian economist, former president of the *Agenzia per il Terzo Settore* (Agency for the Third Sector), recognised worldwide for his studies on civil economy. Since 2019, he has also been president of the Pontifical Academy of Social Sciences. He graduated from the Catholic University of the Sacred Heart in Milan and specialised at Oxford University. He teaches in Parma, at the Bocconi University in Milan and is currently full professor of *Political Economy* at the University of Bologna and Adjunct Professor of *International Political Economy* at the Johns Hopkins University, Bologna Center in addition to being a visiting professor at the Sophia University Institute in Loppiano.

His scientific reflection on the civil economy is recognised around the world and is also the reason why, after a dialogue that began a few years ago, we invited him to do this interview.

Why is an architect interviewing an economist to talk about cities?

D. Costi and G. Leoni (eds.), *Smart City: A Critical Assessment*, The City Project 6,
https://doi.org/10.1007/978-3-031-51288-9_3

Certainly because of the content of his reflection on the civil economy, his interest in the *Smart City* and the impact that the fourth industrial revolution is having on society in terms of work. But also because he has always been interested in, studied and worked on the themes of the third sector and sustainability, essential themes of contemporary society, and certainly because there is also a reflection on the theme of spirituality that interests us a great deal.

There is also a reason linked to the structure of our research: today we are launching a series of interviews that we will conduct with international experts on the themes regarding the city developed by the *Smart City 4.0 Sustainable Lab*, which involves the University of Parma, the University of Reggio Emilia and Modena, the University of Bologna and the University of Ferrara. The research activity started with ten PhD students from these universities who began a critical reflection on the themes of the contemporary city in an attempt to bring together the disciplines that are necessary today for its transformation.

The first step was a critical reflection on the scientific literature on the themes of the contemporary city, which prompted the PhD students to write a glossary encompassing the many definitions provided around the world for concepts such as *Smart City*, *culture of technology*, etc. through a comparative examination of these themes. The interdisciplinary profile of these ten PhD students is the same profile that we wanted to revive by interviewing important professionals, outside the disciplinary sphere of architecture, but who are in fact essential because they have always been a catalyst for our reflections on the city, including design.

As part of our research, today's interview starts with three general questions that we will also ask the other interviewees: one on the culture of technology, one on the *Smart City* concept and one on the topic of the *Wise Town*. This will be followed by a series of questions reflecting on the scientific work of Stefano Zamagni with respect to the themes of the contemporary city.

The first question concerns the word technology, which originated as an expression of the human spirit and, according to the Treccani encyclopaedia, is the practical ability to work towards a given end. We see the analytical and dialectical construction of a "culture of technology" as an ethical necessity and a central issue in contemporary thinking about the city, but not only.

With each industrial revolution, critical thought has debated the risks of possible subordination of human beings to technical tools. Before us, intellectuals such as Walter Benjamin, artists like Paul Klee, theologians like Romano Guardini and architects like Mies van der Rohe questioned the dangers of the industrial revolution they were living through. In the current situation, an immanent critique of what Arjun Appadurai has called an "increasingly pervasive dependence of the world we inhabit on digital networks and mobile technologies (as well as the opaque structures underpinning those systems)" is becoming increasingly urgent. In fact, the development of ICT (Information and Communication Technologies) has accentuated an ever-increasing delegation of control to technological devices, creating mechanisms that proceed regardless of the wishes of individuals. How can we now imagine a "more thoughtful, more human technology, a spiritualised science", as Romano Guardini himself put it, that places people at the centre, developing the skills and tools necessary to enable individuals and communities to reclaim knowledge and expertise?

STEFANO ZAMAGNI Thank you, that is a critical question, which I will answer in the following terms.

We are dealing today, in the age of the so-called digital revolution, with two paradigms, or rather two projects. On the one hand, there is the transhumanist project and on the other, the neo-humanist project.

The former has its stronghold in America, specifically in California, where some 15 years ago a new university with a revealing title was founded: *The University of Singularity*. The goal pursued by the transhumanist project, whose leading exponent is the scientist Raymond Kurzweil, is to achieve complete replacement of man by machine by 2050, i.e. in 30 years' time. In other words, the progress that has been made in the last 20 years on various types of artificial intelligence, as well as the converging technologies of the NBIC (*Nanotechnology Biotechnology Information technology and Cognitive science*) group, allow transhumanists to maintain that the time is not far off when man will continue to exist but will be totally irrelevant. The very meaning of the term transhumanist—*trans* in Latin means beyond, i.e. beyond man—implies an advancement towards the liberation of man from a whole series of needs. I will quote one: Raymond Kurzweil has promised that by 2050 the average life span of people will reach 120 years through the progressive replacement over time of organs that cause physical decay and thus death.

On the other hand, there is the neo humanist project, which is based primarily in Europe because humanism is a European phenomenon, or rather an Italian—or better still, a Florentine and Tuscan—phenomenon. The cornerstone of this project is to avoid *digital servitude*. Machines, artificial intelligence and robots are things to be appreciated and valued, but they must be complementary and not substitutes for the human. Of course, new technologies must remain a tool and cannot become a reason for living. Time will eventually tell whether one or the other project will prevail; what I must note with great disappointment is that while the Americans and the Chinese are investing huge resources in the transhumanist project, the same is not happening in Europe for the neo-humanist project. This is a major shortcoming for us as Italians and Europeans. We already know how little Italy dedicates to scientific research, in terms of both monetary resources and attention. However, the European Union is not pursuing this objective either. It is clear that, apart from a few individual voices, there is no awareness of these issues at institutional level.

That is why we must raise our voices, lest the Europeans decide to go along with the other project and thus accept that it is technology that will guide their consciences. I do not share this view.

D.C. Neither do we, as is clear from the tone of the following questions.

The word *smart* is at the centre of the cultural debate. The term encompasses various meanings ranging from intelligent to lively, but also sharp and fast. In the urban context, the neologism *Smart City* has been coined in specific reference to the city. In *smart* models, we can see a number of limitations that stem from a reduction of territorial governance to its technical dimension, the outsourcing and privatisation of public services and the *digital divide*. How can intelligence be applied to the city in order to develop citizenship rights and reduce inequalities?

S.Z. I am well aware that the expression *Smart City* has now entered common parlance, but it is a dangerous and fundamentally wrong expression. I am amazed that certain intellectuals use it without thinking about the proper meaning of the term. As long as we are talking about *Smart Factories*, the expression may even be acceptable. However, the idea that the city must be *smart* is really "the night of thinking".

We know that Italy has given the world a model of civilisation known as the city-based civilisation model. The city is an Italian concept of which we should not be ashamed; other peoples and cultures never possessed the original concept of the city, although everyone uses it today. This model first appeared in the twelfth century and reached its zenith in the Renaissance, when the Renaissance city took shape. The epitome of the Renaissance city is the "City as a work" where "work" has a very strong meaning implying a city understood not only as the place of production, which is the mistake made by those who speak of *Smart Cities*. Human beings also need to cultivate cultural values and especially spiritual values. The *Smart City* does not deal with spirituality, which is why I prefer to use the term *Care City*. In the "Care City", the focus is not only on efficiency and productivity. The city has to go back to the Renaissance model, to the time when the first hospitals, hospices, and confraternities emerged. Before that, the sick would be left to die in the woods, devoured by wild animals. It was in the twelfth century that institutions such as the *Misericordie* emerged, voluntary associations founded in Tuscany and so called because they set up the first hospices for the terminally ill.

We must be careful with the use of words, and *Smart City* is a term that must be abandoned as wrong, unless we want to agree with the transhumanists. It saddens me that too often we fall into this inconsistency. On the one hand, we want to preserve the primacy of the human being and, on the other hand, we talk about *Smart Cities*.

It would be better to talk about *Care Cities*.

D.C. As a dialectical and cultural counterpoint to the often simplified and trivialised concept of the *Smart City*, we are exploring the idea of the *Wise Town*. What alternative approach to technology can develop a city that is not only fast, quick and connected but also more human-friendly, wise, equitable and inclusive?

S.Z. The concept of *Wise town* is already better and is a step in the direction I mentioned earlier.

Wisdom is not the same thing as intelligence and the two terms do not always coincide, as the Greeks taught us when they used the word *phronesis*. Talking about a wise city is better than talking about a smart city.

To clarify the point, we should recall what Cicero taught us about the distinction between *civitas* and *urbs* some two thousand years ago. The *civitas* is the city of souls while the *urbs* is the city of stones.

The *urbs* model sees the city as a space, whereas the *civitas* model sees the city as a place. While space is a geographical category measured by quantitative variables, place, on the other hand, refers to a living environment.

This is why it is important to speak of the wise city, because it means reclaiming that ancient distinction that the Romans had already drawn. The problem is that if we look back at history—as architects are well aware—we have transformed the *civitas* model into the *urbs* model, and now we are paying the price.

Returning to these roots, typical of our cultural matrix, is a hugely important enterprise at this historical moment. We need to go back to thinking about the city as a place of organised coexistence. The issue is at the heart of Book XIX of the *City of God*, where Augustine revisits the comparison with Cicero. The bishop of Hippo makes it very clear that the city is a public place inhabited by a people. But what makes it possible for different people to live together, transforming a shapeless aggregate of individuals into a true people? Compared to the Ciceronian answer, which evokes the primacy of justice, Augustine identifies *concord* as the glue that creates a people: not only does love (*cum-corde*, putting hearts together) not weaken institutions or destabilise the political sphere, but actually represents its true generative principle.

D.C. Let's delve a bit deeper into your studies to try to understand which of the concepts, on which you have been working for years, might be useful for the contemporary city.

With your definition of "Civil Economy", you discovered a series of moments in the history of the Christian West that testify to the historical depth of that conception of economic life linked to the "principle of reciprocity". What contribution can this tradition of economic and social culture make to focusing on the "culture of technology" we want to discuss and on the *Smart City* that becomes more of a *Wise Town*?

S.Z. Political Economy and Civil Economy are two paradigms that compete for attention in today's world. Until a few decades ago, the Civil Economy paradigm was unknown. I discovered it almost by chance 27 years ago. Until then, it was not even mentioned in economics.

While the Political Economy paradigm originates in Scotland, apparently the brainchild of the great economist and philosopher Adam Smith, the Civil Economy paradigm emerged in Naples in 1753 when the Federico II University established the world's first chair in economics; the chair was named the Chair of Civil Economy and the first professor was Antonio Genovesi, Abbot of Salerno. There are many points of overlap and contact between the two paradigms, but there are also many differences, and one of these lies in the anthropological assumption.

The Political Economy paradigm is based on the anthropological assumption of *homo oeconomicus*, an expression that in turn derives from the concept of *homo homini lupus* discussed by Thomas Hobbes in 1651, according to which every man is a wolf to another man. Political economists accept this assumption and see the market as the most effective way to control the voracity of hungry wolves.

The Civil Economy paradigm, on the other hand, is based on the anthropological assumption *Homo homini natura amicus*, according to which every man is by nature a friend of another man. We can immediately see the difference between the two assumptions: the positive view of man by the civil economist and the negative view of man typical of the political economist, who sees the city as a place of defence against external attacks. Instead, the civil economy perspective sees the city as a place of social and civic friendship, in which man, aware that every man is by nature a friend of another man, will build the city in a way that emphasises the bond of social friendship.

That is why these concepts must be specified when talking about the cities of the future. Everyone is free to embrace the perspective they prefer. However, they must declare it, which many town planners and architects do not do. If city architects assume that men are wolves, it is no wonder that they try to design the city accordingly. Misunderstandings arise when people do not have the courage to clearly state the starting point of their discourse.

I will close by saying that the word paradigm is a Greek word meaning gaze, the way I observe reality. The civil economist observes reality through a different lens than the political economist. We are free to choose either point of view, but we should specify it openly so that "the man in the street" knows where to direct his support. In the absence of clarity, we run the risk of mystification.

D.C. So can the value categories of the Civil Economy—reciprocity, gratuitousness, fraternity—be somehow extended to the contemporary settlement? How can they help us shape the city of the Fourth Industrial Revolution?

S.Z. For the Civil Economy paradigm, the fundamental concept is reciprocity, while for Political Economy it is the exchange of equivalents. The latter is the principle whereby the giver (or doer) has the power to demand payment of a price that constitutes the equivalent of value. Reciprocity, on the other hand, is giving without losing and taking without depriving!

The most immediate example of a relationship of reciprocity, which is the incarnation the principle of fraternity, is that of intra-family relationships. The relationship between a parent and a child is one of reciprocity in which no price is established for what is done or given. The person who makes the first move expects that the gesture will be reciprocated in the future when they find themselves in a situation of need. Those who do not keep the commitment to reciprocate are considered opportunists. Such people certainly exist but they are no more than 10–15% of the population. Furthermore, most people reciprocate the free gift according to the principle of proportionality, whereby the value of the good given and that of the good reciprocated may also be different. This depends on the circumstances and the capacity of those involved. Conversely, in the case of exchange, the rule of equivalence of value applies.

That is why thinking of the city from one or the other perspective makes a huge difference. I believe that we cannot envisage a human-friendly city, a care city or a wise city if we allow all of the space to be occupied by the logic of exchange. Exchange is necessary, of course, but it must not dominate within the care city.

D.C. You spoke recently about an interesting phenomenon that has emerged in recent years: "Strategic Philanthropy" as the recent stabilisation within a permanent framework of "Corporate Social Responsibility". Until a few years ago, this was implemented in a less systematic manner: consider, in addition to the prominence of the Third Sector, the growing spread of B Corporations, and the associations of businesses and citizens that have grown exponentially in number and activity in recent times. This extraordinary opportunity sees the new resolve of the private sector contrasted with the difficulties of a public sector reaction, with Administrations largely unprepared and lagging behind on many issues such as sustainability, strategic vision, openness to dialogue, planning capacity, and initiative in political action. Are we in fact returning to the era of the Prince? How can the Public sector make up for this distance and this delay?

S.Z. It is more urgent than ever to distinguish the two main forms of exercising authority: the form of *government* and the form of *governance*. Note that in Italian we have to use the English terms because Italian has only one word (*governo*) for both.

In English, *government* means government in the strict sense, while *governance* indicates the ways in which decisions taken by the government—that is, by those who exercise formal authority—can be implemented.

The trouble with the Italian situation, which came to light especially during the pandemic, is caused by confusing the two concepts. If we think that the national, regional or municipal government, which has the task of making decisions, is also responsible for implementing them, we are falling into serious error. The government must make the decisions, but the management of the implementation of those decisions must be entrusted to the so-called intermediate bodies of society, namely those subjects that identify neither with the public nor the private sphere, but rather with the sphere of organised civil society.

We need to break out of the public–private dichotomy by entrusting very specific co-planning and co-programming tasks to civil society. We must free ourselves from the force cage of the dichotomous state-market paradigm to embrace a state-market-community trichotomous model.

Our constitution already provided for this scheme in Articles 42 and 43. Article 42 speaks of public and private while Article 43 speaks of community, i.e. communities taking care of certain services necessary to enforce and implement decisions taken at government level.

We cannot continue thinking that areas not covered by the private sector must necessarily be handled by the public sector: people who think along these lines show a certain mental obsolescence. In some cases, the Public sector must intervene, but in others the community must take action. At the city level, *governance* cannot be entrusted solely to the mayor or the city council, who have the final decision. Once the final decision has been made, implementation must be joint, as stipulated in Law 329 of 2000, which was a great law but unfortunately never implemented.

D.C. You have written about the impact of the Fourth Industrial Revolution on society. What risks and opportunities do you foresee?

S.Z. I see two main risks.

The first is the gradual disappearance of critical thinking. One cannot argue or debate with machines, even the most intelligent ones, or with technology, whereas one can argue with science. This risk was the central concern of the recently deceased philosopher Emanuele Severino, who saw before many others that the domain of today is not science but technology. Just think of how technology is used today, where the procedure is imposed by the algorithm. The second risk is the progressive scarcity of relational gods. A characteristic of such goods is that they require an investment of time for their production and generation. A typical example of a relational good is friendship, which cannot be cultivated by remote contact alone. The risk associated with the mass introduction of new digital technologies is that they will consume the time necessary for this purpose. From this perspective, *social networks* are the real danger, because they can trick people into believing they are a substitute for relational goods. They confuse relationship with contact: social networks enable contact but not relationship, which presupposes physical proximity. Since relational goods are the key ingredient to satisfy the need for happiness in every human being, it is clear that the prospect of a gradual decrease in relational goods poses a serious problem.

D.C. You then focused on the highly topical meaning of the word "Responsibility" as an antibody to be strengthened against the risks of replacement posed by "*Smart Machines*" when they are able to "think" and make decisions. What does it mean to be "responsible" when it comes to avoiding this danger?

S.Z. There are two concepts of responsibility that must be carefully distinguished. The first is related to the Latin word *respondeo*, which means to answer. A responsible person is one who knows how to answer for the actions he performs. In recent literature, particularly legal literature, responsibility is understood as imputability. The Americans use the term *accountability*. In other words, you are responsible for what you do and you pay for the consequences if they are negative.

The other notion, derives the concept of responsibility from the Latin *respondus,* i.e. the burden of things. According to this meaning, a responsible person is someone who shoulders the burden of things. In other words, someone who takes care of the individual or collective needs of the environment in which he acts. While in the first meaning one is responsible for what one does, in the second meaning one is responsible for what one does not do although one could do it.

In other words, the sin of omission is always more serious than the error. The hypocrisy of so many people lies in their fear of making mistakes, which makes them passive spectators. Instead, what is needed is to get involved personally, to get one's hands dirty and try to do the good that each of us is capable of, instead of doing nothing, in the name of a kind of moralistic narcissism that allows no room for mistakes.

At this historic moment, it is necessary above all to be responsible towards the environment by striving to achieve the 17 goals of the UN agenda. Recovering the dimension of responsibility towards individual and collective needs is the most pressing and relevant need today.

D.C. For some years now, you have been President of the Pontifical Academy of Social Sciences. Together we started to think about the *Challenge of Urban Cultures* contained in the Apostolic Exhortation *Evangelii Gaudium (The Joy of the Gospel).* On that occasion, Pope Francis suggested tapping into the existing spiritual dimension of the periphery to be interpreted and appreciated for its spontaneity. Will there be room for the "Spirit" in the Smart City?

S.Z. No, there cannot be any room for the spirit in the *Smart City*, but there is in the *Wise City* or the *Care City.* As we said previously, the *Smart City* corresponds to a need for efficiency.

In American economist Edward Glaeser's latest book *Triumph of the City: How Urban Spaces Make Us Human*, the author, in the context of his cultural background, explains how the problem of humanity is simply one of efficiency.

Efficiency is an important value, but so is spirit. This awareness must lead us to proactive action. For example, we must actually create *City Schools*, just as *Business Schools* were created in the last century, i.e. schools where people are taught to manage the city as a common good. As long as these issues are not taught, they will not be accepted.

We should not imagine that the things I have mentioned in this short interview can come to fruition and spread on their own without reflection and teaching. This could be a concrete example to kick-start an alternative project to the one that has been implemented so far. A city that deludes itself into "selling itself" as a space exclusively for consumption, a container for individualistic impulses, would contribute nothing to the most generative social formations, such as the family, or to the great battles for rights, global health and social justice.

D.C. We plan to provide training for administrations as well as for professionals and students, and we will definitely take the name you suggested as a cue and request your involvement. It is a key issue for us as well to bring the public administration together to share in critical thinking on these themes: from the technological aspects of efficiency to the spiritual and cultural aspects.

Thank you for a truly inspiring contribution.

S.Z. Thank you, I wish you success in your work.

Conversation with the Sociologist Derrick De Kerckhove

Introduced by Dario Costi, interviewed by Andrea Fanfoni
2 November 2020

DARIO COSTI Derrick de Kerckhove is a Belgian sociologist, naturalized Canadian. He is considered one of the world's leading experts on new media. Formerly a full professor at the University of Toronto in Canada and at the Federico II University in Naples, now teaching anthropology of communications at the Polytechnic Institute of Milan and also scientific director of Media Duemila and Osservatorio TuttiMedia. After his studies with Marshall McLuhan, whose spiritual heir he is considered to be, he undertook in-depth research on the media's ability to influence human perceptual capacity, starting from the assumption that mass media can be considered as psycho-technologies, that is, technologies that address and affect people's mental activities. This is the reason we invited him to do this interview because as part of the Smart City Regional Research Lab we have been wondering about what we can make of the terms in the current debate. The contemporary city is confronted with the great transition of the fourth industrial revolution, and the Smart city scenario opens up the fundamental question of the impact and influence of interactivity and interconnectedness on our daily lives and social mores. It is therefore crucial to better understand what are the dynamics that are affecting the way we are together. The professor has written some very important books on these issues: back in the 1990s *Brainframes* and *The Skin of culture*, and he came up with the concept of "connected intelligence," taking further the French philosopher Pierre Levy's definition of "collective intelligence" to add a dimension involving individual connectivity rather than only a generalization. We as architects and city scholars need his help to focus on some of the central concepts of our design thinking with the aim of providing an overall reasoning to the Emilia-Romagna Region for the elaboration of guidelines for the contemporary city.

ANDREA FANFONI Let's start with three questions that emerge from a multidisciplinary reflection on the contemporary city. The first question is about the word technique.

Technique originates as an expression of the human spirit and according to the Treccani encyclopedia it is the practical ability to achieve a given end. We think that the analytical and dialectical construction of a "culture of technique" is a central issue in contemporary thinking, and not only about the city.

With each industrial revolution, critical thinking has questioned the risks of a possible relationship of subordination of humans to machines. Before us, intellectuals such as Walter Benjamin, artists such as Paul Klee, theologians such as Romano Guardini, and architects such as Mies van der Rohe have questioned the risks involved inrh the rise of modernity. In today's situation, the immanent critique of what Arjun Appadurai has called an "increasingly pervasive dependence of the world we live in on digital networks and mobile technologies (as well as the opaque structures on which those systems rest)" becomes increasingly urgent. Indeed, the development of ICT (Information and Communication Technologies) has accentuated an increasing delegation of control to technological devices, realizing mechanisms that proceed independently of the will of individuals. How can we now think of a more thoughtful "culture of technology" that puts people at the center, developing the skills and tools necessary to enable a reappropriation of knowledge and expertise by individuals and communities?

DERRICK DE KERCKHOVE This question takes us back to times that were not defined by technology. I was born during that time and twenty years later, technology was still taken for granted as part of a normal development of 'progress' along with many other factors, all 'technological', of course, such better automobiles, radios, medical innovations and scientific discoveries, or the atom bomb, but only a few thinkers, such as Leroy-Gourhan, Giedion, Mumford or McLuhan had isolated technology as a driver of social change. Today, society is beginning to get the message. Technology is what is happening to us, and it is here to stay. The future is even more complex: we, the people, are becoming technologized.

The first industrial revolutions were related to the mechanical extensions of the physical body, emerging as a myth in Pinocchio or as satire in Charlie Chaplin's Modern Times, both implicitly critical of mechanization, the first in a pedagogical key, to recover the human dimension beyond the technology (the whale matrix spits out the puppet in a human body), the second as a forewarning of dehumanization (the assembly-line worker becomes part of the assembly). But the revolution we are living today is addressing not only the body but the mind. It is a cognitive, software revolution, called the "digital transformation". Spurred by AI and ML (Machine-Language) software, the industrial project is to endow machines with enough cognitive skills to eventually run by themselves, a trend that many people fear will replace humans. A good question, however is whether the jobs that are taken over by machines to do the "hard work" were ever 'good' for labourers in the first place. Industry is blindly pushing technology forward without asking questions, but it has a good nose, finding ways to do things better, faster and pour products on the market, turning labourers into consumers. That trend, in effect, is indeed putting people at the centre. It's called CRM or client relationship management. Turning workers into consumers seems to be a good thing but it creates two new problems, the disposal of invasive waste, and climate change with raging environmental damage. Disney's interpretation of Pinocchio saw that coming when during the absence of Gepetto, the puppet uses the magic broom that starts the clean-up but, because Pinocchio does not know how to stop it, floods the house. That's where we are now. Even as the fourth industrial revolution is delegating operations to AI and producing technologies that allow users to delegate not only physical but also mental labour, such as remembering, thinking, and choosing, we are reminded of "la chanson du décervelage" the 'debraining' song created by Alfred Jarry to criticize the stupidity of his professor and of the education system during the 1880s, that is, about the same time as Collodi created Pinocchio (1881). Things have hardly changed since in that domain. Education today is still principal among many things that unprepares humanity to deal sanely and critically with its dependance on technology. As for the 'reappropriation of knowledge', good luck with that in the coming era of language modelling and chatGPT!

And the problem that people's dependence on digital technology is provoking is much deeper than relying on machines to transport, clothe, feed or heat us. Newer machines are also replacing literacy (not to say language itself) as the main social and cultural operating system, by introducing algorithms, which are faster, more precise, comprehensive, and efficient, taking account of infinitely more parameters to arrive at decisions and, in minor instances, their implementation. Until industry and business began to delegate command operations to algorithmic sequences, the main operating system of western society was alphabetic literacy. Applications such as code of law, archive of precedents and history, operating manuals, scientific treatises and discoveries, health records and prescriptions, teaching material and fiction, also educational, all was predicated on a little series of more or less thirty signs (with local variations) that people decoded to go about their business, or wrote to remember what they had done, what they should do or hope for, not to ignore communicating with each other. None of this is gone, of course, and it will be with us for a very long time yet, but it about to take second place in human affairs, even as main decisions in judicial, health, finances, strategy and employment are being

Fig. 1 It's called reading. It's the way people install a new software in their brain

taken by machines. Humans are still, for a (short) while at the helm, but they are not in the the loop! The new operating system of humanity, as a stunned mother observed in the behaviour of her one-year old daughter with an iPad, is digital. Even making up one's mind about anything in news of any kind in our day-to-day life has become hasardous as, in the cacophony of fake news and alternate truth, language has ceased to be trustworthy or relevant. Our senses, too are no longer needed to verify virtual experiences as the algorithm takes their place. The algorithm needs order but not sensory experience. By going online or in the metaverse, the human mind and body, as a result, have experiences that affect them sensorially and intellectually, and with each revolution they face different challenges (Fig. 1).

A.F. The word *smart* is at the center of the cultural debate.

The sense of the term accommodates different meanings that range from clever to lively to sharp and quick. In the urban context, the neologism *Smart City* has been coined as applied to the city. In smart models we see a number of limitations that come from a reduction of territorial governance to its technical dimension, the outsourcing and privatization of public services, and the digital divide.

How can intelligence for the city be declined in order to develop citizenship rights and reduce inequality?

D. De K. Over the last few years, and especially during 2020, urban matters have been changing very fast, for the worst in Putin's war where towns and cities are the object of destruction, rather than just 'the enemy' (involving civil society even more than the military), and for the better going beyond the merely technological control to the human element in the rising concept of digitally 'twinning' the city as in Singapore, Helsinki or Amaravati in India. Twinning the city involves representing it in full 3D simulation and tracking how it is being occupied by the inhabitants, so as to make the urban environment safer and more livable for everyone. To help pushing that trend even further, taking advantage of the rediscovery of Neal Stephenson's 'metaverse' (Snow Crash, 1992)[1] by Meta, I imagined something different from Mark Zuckerberg's project of yet another way to make private money. By the word 'metacity' I intend putting together the technological underpinnings of smart, twin and meta technologies into a single digital public service destined to serve digital—and by extension, also pysical—citizens. The idea being to add to the smartness of technological tracking, and to the virtual human presence in the city, the actual, that is, active presence of the citizens in digital interactions. To allow direct consulting with the township authorities, the metaverse brings citizens what is missing in smart

[1] Neal Stephenson, *Snow Crash*, 1992.

and twin, that is the possibilty to 'occupy' the digital city and interact with each other as if in the 'real' urban environment.

Regarding the issue of citizenship, I am part of a founding group that created the *Manifesto of digital citizenship*[2] a few years ago (January 2018). The metacity concept truly confers citizenship in the virtual as well as the physical reality. With my students at the Milan Polytechnic, where I teach a course in communication anthropology, we decided to change the ground of the cultural, social, technological and political experience of our world. By providing people with a new way to occupy their urban settings and give them the power to access the communal authorities, it is possible to change the basis of social and political organization. In a'thought' experiment, I asked a group of students to study how to combine the Ministry of Environment and the Ministry of Health to crossfertilize the cultural and institutional responsibilities of the two ministries. Among the objectives of the metacity is to reduce the problem of environmental ignorance and neglect, and also to provide citizen with a new image of where they live, not just of their immediate neighbourhood but a fuller overview of all its components. I believe that the intelligence of the city comes, first of all, from the feeling of belonging and pride of its citizens. The cities of southern Italy are not proud of themselves, and it is evident from the state of neglect of those places. Italian cities used to compete against each other to be more beautiful, but today this healthy competition no longer exists. We need the intelligence that comes from the heart of the city and not merely from technology.

A.F. In this regard as a dialectical and cultural counterpoint to the often simplified and trivialized concept of Smart City we are thinking about the idea of Wise Town. What alternative approach to technology can develop a city that is not only fast, rapid and connected but also more human-scale, wise, equitable and inclusive?

D. De K. From what precedes, you can understand why I like the idea of the Wise Town. It is an imaginable concept that could really come to fruition as well as that of the *Proud* Town. People need to recover the pride of their city. The first lockdown caused by the Covid-19 outbreak was a lesson in wisdom. We glimpsed the benefits that can emerge from a wise town. Eliminated for a while were the excesses of traffic, noise and consumption but as soon as the forced confinement ended everyone forgot all about it. Governments were not smart enough to take advantage of that lesson and better regulate for cleaning-up and rationning allowance for motorized traffic, for example by implementing Carlos Moreno's idea of the 15-min city.[3] So, we are back where we started. It is hard to convey to governments and to the people how easy and necessary it would be to make a city happy. Living between Rome and Vico Equense across the Bay of Naples, I would often drive by Torre Annunziata, a city once the pride of Campania, not just for being among the principal producers and exporters of locally famous pasta, but also a prized resort town for wealthy Neapolitans, competing with equally prosperous Pozzuolo on the other side of the bay. Torre A. was terribly run down by mafia control and communal indifference and my heart sank every time I had to cross it. So one day, as my friend, Dr Antonio Irlando became a councillor for culture, we created the concept of the *felicità urbana* ('urban happiness'). The project aimed to stimulate citizens to take action so as to create or regain a lost balance. The idea stemmed from the need to restore hope to those who live in areas where malaise and discomfort have taken over, or support development that had already begun, for example in Fiuggi, southeast of Rome. In Torre, the project was meant to involve first school children getting to know and describe the namesakes of the streets they lived in, but all citizens were to participate in the process of helping their children and friends to provide a

[2] Massimo Di Felice, *Digital citizenship and the end of an idea of world*, Journal of E-Learning and Knowledge Society, 18(3), 22–28, 2022.

[3] Carlos Moreno, *The 15-min city*, Ted Talk, https://www.ted.com/talks/carlos_moreno_the_15_minute_city/transcript?language=en.

Fig. 2 In Fiuggi, Piero Fantastichini, avant-garde artist, exposed his Tower of Happiness project composed of tiles created by the Fiuggi school children

deep description of the town. What always worked well to start urban happiness projects was a method that I created and called *Connected Intelligence workshops*. For example, in Fiuggi, we brought together school children with Piero Fantastichini, a well-known artist, to create a public sculpture constituted of individual tiles made by the children (Fig. 2). Workshops were conducted and projects presented also to Torre, to Gela, to Potenza, to Soveria Manelli, to Vico Equense and also to rescue the Sarno river from pollution. The workshops always generated interest and enthusiastic adoption of the projects initially but would eventually lacking follow-up that ensures real change. The ensuing indifference of the communal authorities and lack of minimal funds would systematically bring the project to a protracted end.

A.F. Indeed, starting from the concept of collective intelligence you coined the term connective intelligence that is, the idea of intelligences interacting individually within the network. Then you talk about connective architecture, an architecture that provides for the physical and mental interconnectivity of bodies and minds. So I ask you if we can also talk about connective cities and what characteristics should they have?

D. De K. The example above illustrates both my understanding of connected intelligence and urban connectivity. Each tile would bear a Qr code linked to images and text telling the story of their street's namesake, readable not only by tourists and visitors in front of the monument on their smartphone but also by anybody anywhere in the world, of interest, among others, to expatriates wishing to reconnect with their place of origin. The other benefit was to increase the sense of pride of the local citizens becoming aware of the fact that their little provincial abode had suddenly taken on a global dimension.

The idea of a smart city is that of an interconnected city where everything converges, where people are entirely connected. The future of humanity on earth is represented by this electrical interaction that now goes through a *digital* phase, but tomorrow will feature the *quantum* phase correlating all the IoT (Internet of Things, made up of sensors everywhere), the IoB (Internet of Behaviors, connecting every move made by everybody), the IoP (Internet of People, bringing potentially everyone in contact with everyone else). This means that nothing can escape this bond of connectedness. This electronic straightjacket condition obviously creates difficulties on the level of individual identities and the autonomy of the person. That has to be negotiated in a culture such as that of the West where individuality is prized above all else (in education, sport, business and employment) and privacy, albeit in sharp decline, is still treasured. People have to recognize that digital interaction brings them to total interconnection, not only in the smart city, but even more so in the metaverse where eye movements need to be recorded in order to be functional. Can we afford to lose our civil autonomy and freedom of conscience that we gained with such strife, torture and bloodshed during the Renaissance? Today mental, physical and social independence is thrusted inside personal portable devices. The alternative is to change the ground, to accept connectedness while maintaining personal values. Ethics lies in figuring out how to harmonize the individual's presence in the world, and how to be useful in the city. The great "sick people" of our time are those who have not understood this concept and aim only at economic profit. They are sick, they are not bad. The big American capitalist companies make billions in revenue without knowing how to spend it except on themselves. But the extraordinary architectural and cultural heritage of Italy comes from rich people who were proud of their cities and competed to present the best accomplishments of urbansim and art.

A.F. Picking up on your words, you speak of the shift from the Renaissance "point of view" to the contemporary "point of being" that is, the physical sensation of our presence in the world. Do you believe that the rational organization of the physical space of the urban environments in which we live can be disrupted by the spatial and cultural revolution to which we are subjected?

D. De K. This is a complex question. The shift from "point of view" to "point of being" is not just a sensory issue although the senses are fundamental to understanding it. The issue is the transition from the spectator's to the participant's experience of space. Oddly enough the metaverse concept since at the least the short-lived success of *Second Life*, offers both as you can in your own mind choose to see yourself as an actor on the virtual stage, or see that stage at eye-level. But neither in your mind or even less so in virtual environments are you invited to experience those spaces physically. The "point of being" requires a sensory dimension that allows one to believe not only in the things one sees but also in the things one feels. In this context, women are much better than men, they have much stronger tactile and auditory sensitivities, and that is why they will represent the governments of the future. Mostly women saved their country from Covid-19, from Norway to New Zealand via Denmark. The city of the future will have to support a complex sensory dimension, and a determining factor will be intuition, that is, the total perception of the environment and a keen sense of 'body-language' as well as that of the infinite variations in the intervals between bodies. To explore the notion of the 'cities of touch', there are at least two architects I can recommend: the first one is Finnish and his name is Juhani Pallasmaa[4] and he is one of the very few architects who have really tried to think about the architecture of touch. Perhaps recovering Thomas Aquinas' idea that sight is a way of touching, Pallasmaa also imagines the eye as a tactile medium. The other author is Brazilian and her name is Rosane Araujo. I wrote the preface of the French translation of her book, The City is Me[5] where she reviews all manners of self-identification of the individual person from Immanuel Kant, to Sigmund Freud to Ludwig Von Bertalanffy. She concludes this analysis of precedents with her stunning theory of urbanism in which, she says "the city is me": the places in the city that I know are part of my individual being that relate in an ever changing environment. Wonderful! How about that for western—and a bit eastern too—connectivity? These are two authors for students of design and architecture.

A.F. In a context where technological devices are sensory extensions and where cognitive reality is extended by the network environment, can we control the technological devices with which we interact? Philosopher Adriano Fabris points out that there is a risk that our own responsibility for actions not directly produced by us will be

[4] Juhani Pallasmaa, *From Space to Place, Existential experience in architecture*, https://bigsee.eu/juhani-pallasmaa/.

[5] Rosane Araujo, *The city is me*, Chicago University Press, 2013.

diminished. To what extent will we still be responsible for the effects such devices have on the territory?

D. De K. The second question is a clarification of the first but both deserve an answer. We are replacing our memory with that of our smart phones by enthrusting them with our intelligence, our conscious discernment. We are entrusting digital assistants, Alexa, Siri, Bigsby, with the ability to make decisions for us, and this is just the beginning. The digital assistant is a kind of "Trojan horse" for the arrival of the digital twin, which we discuss in *Oltre Orwell. Il gemello digitale* (*Beyond Orwell, the Digital Twin)*, our book with Maria Pia Rossignaud.[6] The *personal* digital twin (PDT) will become our digital double, as if it were a supercharged digital assistant, benefitting from all the data that concerns us personally, such as for example everything that is contained and ajourned permanently in our smartphone, but also from what I have called our 'digital unsconscious' that is everything that the networks and the databases of the world know about us that we either don't know or don't remember. This PDT, having access to many more data and parameters that we, as physical persons, have, will be always in a better position to not only advise us on what to do, but in ever more instances, do it for us.

Instrumental interactivity, produced by physical interaction with devices, gives us an illusion of autonomy and an illusion of choice. But in the future, the issue of responsibility over choices will be fundamental for personal conduct and for the law. Furthermore, with the arrival of the GPT series (Generative Pre-training Transformer—Open AI, 2020), and more so with chatGPT, people are also delegating the responsibility for their own words to AI. How can we imagine, after having delegated our main deliberation and decision-making faculties to a machine, that we are still capable of being accountable for our actions? Ethical and moral responsibility, personal and collective, to ourselves or to the community requires a manner of resistance at all levels.

[6] Derrick De Kerckhove, Maria Pia Rossignaud, *Oltre Orwell. Il gemello digitale*, Castelvecchi, 2020.

A.F. My question now continues your argument. The great development of smart planning models prompts questions about how to interact with information and communication technologies. In your *Beyond Orwell, the Digital Twin*, you stress the need to lay the groundwork for an "ethics of the algorithm."

How necessary is it to regulate the relationship between ethics and technology?

D. De K. Yes, it is, absolutely—and urgently—necessary. It is necessary to understand what ethics means and what power technology has over it. The first psycho-technology was and still is language, but the one that makes a real difference between cultures is writing itself. When the ancient Greeks invented the alphabet they invented an unambiguous system of translating the sound of human words into visible signs. This offered the possibility of doing two things: reading *silently* and *privately*, integrating all language, all its connotations, nuances and possibilities, as a system of production available for personal use. This is the critical origin of the western world's strong bias for individualization where the very notion of the 'person' was created.

The impact of alphabetic technology on humans enabled their formation by integrating all the things they read. By writing a person becomes a producer responsible for a future life path. This means that people who read can internalize language into thinking without speaking and thus become endowed with an inviolable private conscience capable of taking charge—and responsibility—for their own destiny. This is fundamental to understanding what is happening today. Today there is a new dimension of human responsibility that exceeds that of the individual and that of the local community. The responsibility of the future is a collective one to the environment and the earth we inhabit.

Responsibility is always to something or somebody. In so-called 'shame' societies, the responsibility is directed to the 'other', the family, the social group, the clan; in 'guilt' societies, largely individualistic and supported by Christianity, a religion primarily based on the individual person—and probably dependent from its earliest origin on alphabetic literacy—one's first responsibility is to oneself. You are and you *feel* guilty even if no one knows anything about what you did because you know you did it, you know it was wrong and you feel bad mainly because you have failed yourself even if what you did also involved harm or loss to somebody else. The saying 'Love thy neighbour as thyself' puts the self before the neighbour. Shame and guilt are the negative counterparts of responsibility to the other or to the self. Today, however, both shame and guilt need to be complemented the anguish we feel for the fate of all humanity, and beyond, for all living and material being. Anxiety is what corresponds to shame and guilt regarding the environment. We tend to brush it as someone else's responsibility, say the petrol and gas industries, but we all share into the problem and we know it when throw or even see garbage in air, water and land.

A.F. You speak of e-nquisition that is, a process that will unmask our souls through the electronic inspection of our behaviour. The fact that cyberspace is a public space where the individual is ubiquitously subjected to the control of his or her actions undoubtedly presents us with an ethical problem. Does losing the freedom of private identity to Big data alarm you or do you think it will benefit our democratic system?

D. De K. I am not alarmed, just resigned. No one can stop what is going on today. It's a tidal wave that runs over all ideologies and economies. Democracy is the first to drown. It was invented by the Greeks with two Greek words: demos meaning people and kratos meaning power. It is a beautiful idea, a fabulous ideology that we Westerners as their descendants have also more or less learned and followed. But, people hardly believe in voting anymore, as smaller and smaller numbers go to the urns. Other cultures, for example the Chinese culture, have never really experienced democracy. Their cultural operating system doesn't require it. The chinese culture does not come from the same root as ours, although westeners certainly have tried to impose their ideological viewpoints on them. Yes indeed, they have learned the tricks and trade of capitalism, but it doesn't sit well with them. The only ideology that the Chinese imported successfully from the West has been communism, and that, for a good reason, namely that their cultural operating system is ideography, the representation of meaning by visual concepts, not by sound, not by language. This is why the Chinese have never cultivated individualism to the same extent as Westerners. Representing local languages by sound created nationalism. Just consider that from the breakdown of the Roman empire emerged different nations based on local languages. This did not occur in China, even after the breakdown of the Empire because all local languages, including Japanese and Korean, shared the same writing system. It is only after developing their own phonological systems that both Japan and Korea eventually became independent nations based on katekana (twelfth century AD) and hangul (1443 AD).

The easiest way to understand how community is prioritized over individual interests in China today is to examine the government policy of *social credits*, that is, the observation, evaluation, and retribution of suspect businesses and some citizens through sanctions or rewards according to their behavior. It is a form of surveillance being developed with little or no resistance from the people by the government since 2004. Nearly two million Chinese were denied access to some means of transportation because their valuation score was not high enough to be deemed safe travelers or, even worse, because they had too many friends with low scores. Surveillance also happens in the West but for profit rather than to secure benefits for the community at lare. The Social Dilemma,[7] a documentary filmed in 2020 shows how users are being manipulated by the big social networks, Facebook, Twitter, Google, based on an algorithmic profiling that organizes content and orients user tastes and choices from the information provided. Also using profiles, but based on your queries, Google will show different

[7] Jeff Orlowski, Davis Coombe, Vickie Curtis, *The Social Dilemma*, Exposure Labs, Argent Pictures, 2020.

results for asking the same question, in the same language, with the same words, at the same time, by two separate users. Kids' behaviors are studied in the same way Cambridge Analytica studied the behaviors and inclinations of undecided American voters that eventually led to Trump's victory in the U.S. presidential election, or to Brexit in the 'not so United' Kingdom.

I may not agree with all of Xi Jinping's policies, as for example with the Uyghur question, but as Westerners we cannot imagine imposing our cultural system on the whole world. So I am not alarmed by the possibility of losing some individual freedoms even though there will be changes, many of them unpleasant; just think of the deprivation of freedom caused by the lockdown. We will certainly lose the democratic dimension, which we have abused in the past, but, as a counterpoint, algorithmic power will bring some benefits. Since the world is moving from the alphabetic to the digital transition we are facing a profound transformation where technology performs a kind of "possession." Democracy as we knew it no longer exists. So it remains to rethink the way of being social in a world where the private dimension of Greco-Roman culture is over. We have to figure out how to negotiate some benefits that come from the condition of the past such as freedoms, personal ethics, not as a right, because it is already too late to do that, but as a way of being a citizen.

A.F. The digital space can bring citizens closer to participation in public affairs, now strongly viewed with distrust, by fostering an ongoing dialogue with representatives and expanding the moments of consultation. How can new media and the web foster new forms of participation or innovate traditional ones?

D. De K. As I suggest above, the democratic system no longer works; it is gradually giving way to a *datacratic* system. But, to save the illusion of democracy, we must indeed learn how to make the best use of the network resources. For example, in Taiwan the network is used to negotiate some laws, rights and obligations with the people. We should not think that participatory politics is impossible. I propose the model of the metacity precisely to restore participation of the people in the decisions that affect their livelyhood and expectations about their future in the physical city they occupy. I am also in favor of the universal wage, at least to facilitate the transition when machines will have taken over most, if not all human labour.

In such a situation, we have to be concerned about "datacracy" or "algocracy", that is, the power of algorithms. Algorithms don't 'know' anything, they are merely mechanical sequences of command and control operations. Paraphrasing the poet, Alfred, Lord Tennyson, "*Theirs is not to reason why*".[8] Only humans (and other animals) can reason, and that is why they have to participate in decisions proposed, or already too many times *made,* by machines. People's participation, however, can only be relevant if they actually know what they are participating about and into. This requires education, unfortunately, the lowest among most governments priorities. Society in general has abandoned children to an education that is tragically out of touch with their real context through ignorance of technology, indifference to the environment, and the loss of civic values. The digital transition will succeed as long as huge efforts and investments are made at every level of education. With my students at Polytechnic, we considered and discussed the possibility to unite the Ministry of Education with the Ministry of Defense because we thought that in the time of digital culture and the information economy the greatest economic resources should be directed toward education rather than the military. Recent history has tragically proved us wrong. But, in the kind of peacetime we thought we were beginning to experience, the "educated" person is any country's best defense. It is necessary to overcome the stage of incoherence of the human being through the negotiation of a participatory social relationship with individuals intellectually capable of defending these fragile democratic values.

[8] Alfred Lord Tennyson, *The Charge of the Light Brigade*, 1854.

Coming back to the first question, the next 'datacratic' stage could be introduced by quantum computing, that is, by collecting and integrating all the world's relevant parameters. Imagine computing that could consult all of the data, all of the networks, all of the Internet of Things (IoT), all of the time, in real time and distribute optimal solutions everywhere simultaneously. Of course, pertinent access to all the information would be, just like access to the Internet is today, depending on the specifics of the need, some automated, some punctual. Technology has already developed the data language (digits and algorithms) to access and integrate such a massive quantity of information, but not yet the processing power. Quantum physics is still a very complicated science that we know little about, but it promises a radical alternative to the current situation. I don't know when a quantum era will really start but it seems closer every year, not to say month. I can already feel that it will bring a global change of mind just as I felt in the late ninteen eighties that the digital transformation was heralding an exponential change, but, revisited in quantum terms, that hunch intimates a much higher and more profound transformation.

The idea is that quantum physics can overcome the strictly deterministic laws of our Newtonian physics which implicitly have tended to legitimate our detachment from nature, our commitment to technology, our neglect of the environment, and the disturbing situation of disharmony that has led us to the greatest epistemological crisis in history. Maybe the *word* was 'In the beginning', but immersed in the chaos of post-truth, fake news and simulated thinking (e.g., chat GPT and AI generally) one wonders whether it will still be there at the end. A quantum sensibility might make sure it is. An observation worth making is that our technologies keep growing driven by the levels of complexity they inspire in the human mind. The West has long lost its earlier relationship with metaphysics. The average mental, cognitive and perceptual abilities do not access or trust spiritual levels anymore. But a spiritual dimension is neither supernatural nor committed exclusively to religious methods. It also comes quite naturally, sometimes inspired from philosophy, poetry or art (henceforth restricted to privileged elites and turned into exterior signs of wealth). The last truly innovative thinking about metaphysics came from the Surrealists who explored and defined the theory of the 'supreme point', that is, the point that reconciles all opposites, life and death, love and hate, rich and poor. André Breton who was the leader of the movement expressed it in this way: "*Everything leads to believe that there exists a certain point of the spirit where life and death, real and imaginary, past and future, communicable and incommunicable, high and low, cease to be perceived contradictorily*".[9] Therefore it would be futile to look for another motive to the surrealist movement than the hope to determine this point. My intuition, over the last forty years, has been that quantum physics would eventually lead us there, being that approach to reality that grounds it in comprehensive and all-compassing incertitude.

If and when the quantum era begins, there will be a massive return to metaphysics and natural sprituality largely because Heisenberg's 'uncertainty principle', also interpreted as the 'indeterminacy' principle will scientifically legitimate heuristic thinking, based entirely on uncertainty, as opposed to strict determinism.

Conversation with the Autonomous Driving Engineer Alberto Broggi

Introduced by Roberto Menozzi and Francesco Leali, interviewed by Emanuele Ortolan
23 November 2020

ROBERTO MENOZZI Professor Alberto Broggi is one of the leading international figures in the field of computer vision applied to autonomous vehicles. In addition to being a distinguished colleague, he is also long-standing fellow traveller. Since my arrival at the University of Parma, Alberto became the first graduate in Electronic Engineering and together we have witnessed the development of first the Faculty of

[9] André Breton, *Manifestoes of Surrealism*, 1924.

Engineering and then the Department of Information Engineering, which was recently merged with the Department of Engineering and Architecture.

From his degree thesis onwards, under the guidance of Gianni Conte, he worked on issues related to computer vision applied to autonomous vehicles, an activity that led to the foundation of VisLab S.r.l. in the late 1990s, which became a spin-off of the University of Parma in 2005. Thanks to the work and expertise of Professor Broggi and his collaborators, VisLab has established itself as an important global player in this field, to the point of attracting the interest of Silicon Valley companies, which resulted in its acquisition by the American company Ambarella Inc. in 2015. The acquisition provided for the company and the research team to remain in Italy, but also, more importantly, for the development of the Research & Development division with the hiring of additional human resources, which now total more than 50. In early 2018, the VisLab/Ambarella office was opened on the Science and Technology Campus of the University of Parma.

The issues that Professor Broggi deals with are of great importance in the present and future of our society. In fact, efficient and sustainable mobility will be one of the key challenges for the development of our societies in the years to come. It will be very interesting to have the point of view of a researcher, manager and entrepreneur, which Professor Broggi can bring to the topics raised by the questions prepared by the PhD students in the group coordinated by Professor Dario Costi. Among these, I find particularly thought-provoking the reference to the Catholic theologian Romano Guardini, who introduces us to a series of extremely important reflections: as scientists and technologists, we have the task of inventing, developing and making available to society new technologies that are ever more sophisticated and powerful, but as thinking and conscious human beings, we also have the task of understanding how these technologies can lead to a harmonious development of human beings in their environmental context, seen in their entirety and not only as producers or consumers of wealth. Browsing through the questions, I find the juxtaposition of the terms Smart and Wise very interesting: we live in an environment, in a society, in cities that are increasingly smart in the sense that they are increasingly automated, interconnected, informed and communicating. I think it is fair to ask the question whether this development will lead us to increasingly Wise cities and societies or not. The themes that Professor Broggi deals with are certainly a very significant and important part of these major issues that we are addressing.

FRANCESCO LEALI I believe there is much to learn from this dialogue with Professor Broggi and I am most curious to hear directly from him as he answers the very interesting questions he will be asked.

Today we are in a moment of great transformation of everything concerning mobility and everything that is automotive. Professor Broggi had foreseen this well in advance of all those working in this sector. For this reason, he can be defined as the father of these issues, since he started working on them before anyone else. His first results date back several years.

Today, themes related to autonomous driving and the use of electronic and computerised systems aimed at improving driving conditions are a very topical issue. They represent one of the great transformation trends identified internationally, together with sustainability, the ageing population and the progressive urbanisation of cities. Despite a slight countertrend during this pandemic, these are all extremely profound transformations affecting the global system technologically, socially and culturally. I would like to thank Professor Costi, coordinator of the *Smart City 4.0 Sustainable LAB* research lab that we are developing together with other colleagues at the regional level. I would also like to underscore the importance of investigating the depth of the changes taking place and the extent to which they are affecting and will affect all of our daily lives. Today, we believe that the technology is ready and is at an excellent point of development to introduce a certain type of transformation. However, an effort must be made to achieve cultural, social and attitude change towards a type of technology that, as we know, is neither good nor bad by definition, but can simply be used in a positive or negative way, that is, by creating or solving problems or, as is often the case, both. We are talking about technologies that should not only be considered in the automotive sector, but should be seen as part of the broader context of mobility, the city and the environment in which we live. All the technologies we will discuss are unlikely to work unless they are embedded in a digital infrastructure that allows information to be exchanged, processed, and transmitted. Such technologies cannot exist without the ability to exchange information not only with road or city infrastructure but also between the vehicles themselves, opening up many themes of which connectivity is a key aspect.

EMANUELE ORTOLAN Let us start with three questions that emerge from a multidisciplinary reflection and concern three key concepts as a contribution to reflection for research on the contemporary city. The first question concerns technology.

Technology originated as an expression of the human spirit and, according to the Treccani encyclopaedia, is the practical ability to work towards a given end. We see the analytical and dialectical construction of a "culture of technology" as an ethical necessity and a central issue in contemporary thinking about the city, but not only.

With each industrial revolution, critical thought has debated the risks of possible subordination of human beings to technical tools. Before us, intellectuals such as Walter Benjamin, artists like Paul Klee, theologians like Romano Guardini and architects like Mies van der Rohe questioned the dangers of the industrial revolution they were living through. In the current situation, an immanent critique of what Arjun Appadurai has called an "increasingly pervasive dependence of the world we inhabit on digital networks and mobile technologies (as well as the opaque structures underpinning those systems)" is becoming ever more urgent. Indeed, the development of ICT (Information and Communication Technologies) has accentuated an increasing delegation of control to technological devices, creating mechanisms that move ahead regardless of what individuals may want.

How can we envisage "more thoughtful, more humane technology, a spiritualised science", as Romano Guardini himself put it, one that puts people at the centre, developing the skills and tools necessary to enable individuals and communities to re-appropriate knowledge and expertise?

A. B. Perhaps before discussing how to conceive of a more spiritualised science, it would be worth pondering how science and spirituality can coexist. Science and spirituality are two distinctly different concepts. The former is based on numbers, it is deterministic, exact. We could call it demonstrable. The latter, on the other hand, is something more personal, internal and sometimes not demonstrable.

Personally, I struggle to put these two concepts together, which are very lofty, but also quite different, which is why I think they should remain separate.

Increasingly, human dependence on technology is viewed negatively and we often hear talk of the subordination of humanity or slavery to technology and to technical tools that have been invented to meet some particular need. When a new technology is invented, evaluated, modified and then finally implemented, we realise its benefits, and it begins to become so much a part of ourselves that we can no longer do without it.

When we are deprived of something, we immediately miss it, and not only when it comes to the latest technology, i.e. information, mobile and communication technology. A trivial example is heating: in winter, when there is no heating, we are uncomfortable, complications arise and we feel that there is something missing. It is a trivial technology, but one that has now become part of us. As soon as something is invented, it becomes part of us.

Another concept usually associated with technology is that it moves forward regardless of what people want, almost autonomously.

However, technology does not evolve on its own —it needs someone to drive it forward, someone who understands it thoroughly, in depth, in order to be able to govern it. This negative connotation that is often levelled at technology, or at least the need to spiritualise it and bring it close to people, is the result of a poor understanding of technology. Unfamiliarity usually breeds fear and, in this case, an answer should be sought in familiarisation with technology, which may help us gain more confidence. An example I often give is the motor vehicle: when it was first invented, it terrified everyone, it was seen as a diabolical object because it made noise, became hot and moved on its own. This was a view distorted by the fact that people did not know the details of the vehicle or the characteristics of the engine, whereas today this is no longer the case. The more we learn about the details, the less afraid we will be of technology and the better we will be able to handle it.

I actually have a somewhat pragmatic and detached view. I see no difference between a cutting-edge technology such as ours, based on artificial intelligence for driving vehicles, and a whisk for making whipped cream. They are all tools that allow tasks to be performed more easily and faster. Therefore, the message I would like to get across is that we need to familiarise ourselves with the technology, learn how to handle it and benefit from the advantages it can give us.

E.O. As part of this discussion, it would be interesting to talk about the term Smart, which is currently at the centre of the cultural debate. The term encompasses various meanings ranging from intelligent to lively, but also sharp and fast. In the urban context, the neologism Smart City has been coined in specific reference to the city. In Smart models, we can see a number of limitations that stem from a reduction of territorial governance over its technical dimension, the outsourcing and privatisation of public services and the digital divide.

How can intelligence be applied to the city in order to develop citizenship rights and reduce inequalities?

A.B. The Smart City is not merely a settlement that makes massive use of technology without asking how that technology can be managed to avoid creating inequalities and to ensure rights for everyone by limiting the digital divide. In my opinion, the Smart City integrates two aspects: the one related to the substantial presence of technologies and the one that envisages an egalitarian and usable distribution of these resources for everyone. In fact, I think that a high-tech city where no one had access to these resources would be no different from a city without technology.

The advantage becomes apparent as soon as everyone starts accessing and using the technology. Hence the need for a plan to distribute it, in other words, a default deployment managed by a neutral independent party who can steer the use of the technology in the right direction. This is to ensure a fair distribution of technology without it benefiting only those who know it and those who are able to master it. Let me give two examples related to the world of education and healthcare: if a teacher only educated his own children and those who pay him, that would be less than ideal; likewise, if a doctor only treated her family members and those who pay her, we would not have healthcare. On the contrary, we have independent parties who manage the education system and the health system, ensuring access for all. So the key is in the management and distribution of technology.

E.O. You mentioned a neutral independent party who can manage all aspects of a Smart City, who is knowledgeable and able to govern the exploitation of technology in the right direction. As a dialectical and cultural counterweight to the often simplified and trivialised concept of a Smart City, we are looking instead at the notion of a Wise Town.

What alternative approach to technology can develop a city that is not only fast, quick and connected but also more human-friendly, wise, equitable and inclusive?

A.B. Rather than an alternative approach to technology, I would speak of a coordinated approach to technology. Once technology is developed, it needs to be managed, and we need someone to set rules, to define how it is exploited and used. In general, there is no rule to define how to manage technology. It depends on the specificity of the technology. What is important is a detailed knowledge of it, to understand its benefits and the possible problems it may bring. Coordinated action is needed between those who develop and those who manage the technology. The latter must have in-depth knowledge in order to govern it as effectively as possible. Management from a third party perspective, such as an overarching government, must be able to steer it in the best direction. Take the example of the geographical digital divide: it is obvious that a telephone company has more business interest in expanding a high-speed line in Milan than in bringing it to a remote location in the Apennines. That is why we need a neutral independent party and who will lay down rules, for example, taxing the deployment of the telephone line in the city while simultaneously providing a tax break or incentive for extending it to more isolated towns. Without *super partes* oversight, technology will go where the business is and this is not the ideal situation if we care about its fair distribution.

E.O. The Society of Automotive Engineers has codified five levels of automation for cars: from level 0 where there is no automation whatsoever to level 5, where the car is capable of performing every aspect of driving completely autonomously. At what level of autonomy are most of the vehicles on the road today and what improvements in road user safety are we seeing today as a result of the automated processes installed in vehicles?

A.B. This question gives me the opportunity to emphasise a fundamental aspect of the SAE levels, namely their clarity, specificity and definition. Unfortunately, there is a lot of confusion within the media—there are actually advertisements that talk about level 3 being applied to vehicles, but when one looks at the details, it is obvious that it is not level 3. I should point out what the main difference is: the lowest levels from 0 to 2 are the levels where the driver has responsibility for the vehicle, while in the next levels from 3 to 5, responsibility can be delegated to someone else. In particular, 0 has no automation; at level 1, one driving task is automated; at level 2, two or more tasks are automated such as, for example, maintaining the distance from the vehicle in front and simultaneously staying within the carriageway. These are the levels up to 2 and the capabilities of the vehicles we have on the roads today. I personally do not know of any vehicle on the road today with a level above 2. At the moment, there are no level 3—and therefore fully autonomous—vehicles in which the driver can even forget that he or she is inside a vehicle until it encounters a problem or has a particular situation to handle and calls the attention of the driver, who can take back control within a certain time. Normally there is a lot of confusion, because Level 3 is wrongly associated with capabilities of today's vehicles such as lane keeping, maintaining distance from the vehicle in front, and applying emergency braking.

There are also later levels being tested on the road. For example, we are running trial trips between Parma and the centre of Parma in fully automatic mode, but there is always someone who is ready to take control of the vehicle, so we cannot say that these are levels above 3. The chief advantage of these possible vehicle-integrated automations is road safety. In fact, we should remember that 93% of road accidents are due to human causes. Rather than vehicle malfunctions, they are caused by a driver who fails to keep the vehicle on the road, or lacks lucidity because he or she is driving under the influence of drugs or is simply distracted. Replacing the driver with an autonomous driving system would at least mitigate or decrease the number of road accidents. These benefits are accompanied by others, described as side effects, such as improved fuel consumption and optimised journey time.

E.O. Each level of automation corresponds to a greater number of interactions between the vehicle and its surroundings. You have spoken in some of your interviews about the fact that in "simple" contexts, such as motorways, our cars would already be ready to drive in complete autonomy. When will it be possible to achieve full automation in an urban context and what, if any, structural changes will have to be made in these contexts to accommodate these new types of vehicles most effectively?

A.B. Predictions are always difficult to make, especially when looking so far ahead. There are vehicles on the road with advanced capabilities, but they still need someone (safe driver) who is ready to intervene in case of the unexpected. It is difficult to say when full automation will be reached, because full automation means level 5 not level 4 or those below.

Full automation will be achieved when I can locate the vehicle, which will then pick me up in front of my house, I will get in and it will drive me autonomously to the required destination and then park or serve some other user. This scenario will still take years as vehicle innovations take time to be integrated. Today, car manufacturers are testing and integrating current technology into vehicles that will be on the road starting from 2024 to 2025.

As far as dedicated infrastructure is concerned, there are two schools of thought. The first considers infrastructure to be a prerequisite for vehicles, without which the automatic vehicle cannot function. The second believes that the autonomous vehicle must be able to manage itself independently, without any help from the infrastructure. I belong to the second school, the one that sees the autonomous vehicle in every context without reference to other subsystems operated by third parties.

Some mention 5G as a prerequisite for autonomous driving, while others do not agree. Personally, I do not see it as a prerequisite for autonomous driving. Once 5G is up and running all the better, because we will be able to use all the information it can carry. We will have information on the road ahead of what the human eye can see and ahead of what the sensors can detect. It will certainly deliver benefits, but the vehicle must also be able to handle itself safely without 5G and without the dedicated infrastructure. Dedicated infrastructures may have their uses, and once they are in place, they can improve results for system optimisation. However, the vehicle must be able to manage itself independently of its surroundings.

E.O. Autonomous driving will bring about a paradigm shift in mobility: the shift from exclusive vehicle ownership to shared vehicles. This revolution will certainly have implications for the city and the size of the road infrastructure in our urban centres. Combined with the spread of electric mobility, it is estimated that the European Union's 2050 target for an 80% reduction in emissions can be achieved. Viewing vehicular mobility as a need to be fulfilled on demand that is enabled by vehicle automation could lead to an 80% reduction in parking requirements and 80% fewer accidents. A 2019 study by Transport & Environment (TE), Less (cars) is more: how to go from new to sustainable mobility takes a more pessimistic view, arguing that the unregulated expansion of self-driving cars could increase traffic in European cities by 50% to 150% by 2050 because they could become so cheap and easy to use that they would encourage people to travel more often and longer distances by car. The study argues that automation will not solve the traffic and parking problem without proper strategic design planning of urban space and mobility. Therefore I would like to hear your opinion on the urban revolution awaiting our cities and whether you believe that autonomous driving can really bring the expected benefits to our cities and to what extent?

A.B. Both of these scenarios are entirely possible. The first one envisages sharing of vehicles, which will lead to fewer vehicles on the road. The second scenario assumes an affordable cost of the vehicle with the risk of having a large number of self-driving vehicles on the road performing a wide variety of tasks: picking up children from school, collecting a certain product, etc.

Of these two possible scenarios, I personally prefer the first one, the one that involves vehicle sharing while also guaranteeing more efficiency and less pollution. In fact, when we buy a vehicle, we make the least efficient purchase possible, because we buy it and keep it parked and unused 95% of the time. If we could share it with others, the vehicle would always be running, always efficient and the benefit would be tangible. As I said before, these are two different scenarios and both are possible. We need an impartial third party to decide on the direction to take, possibly by favouring the proliferation of shared vehicle fleets to avoid the second scenario. This will benefit mobility and radically change the urban layout. If the first scenario prevails—and therefore the vehicles will no longer be owned but will become shared—they will not need a parking space in front of the house, because they can go and park outside the city by themselves. Parking will be different: there will be no need for car parks where all the vehicles are positioned to leave whenever they want. Instead, if vehicles become shared, they can be lined up one behind the other and, when required, the first one in the queue can leave. The idea of parking will change radically and will also eliminate the personal garage, which will become superfluous with shared vehicles. In addition to the urban organisation of car parks and garages, the urban environment of roads will also change drastically. They will no longer need traffic lights or systems to regulate traffic flow because vehicles will be able to communicate with each other by establishing in advance the right speed to maintain in order to cross paths.

E.O. Since the end of the last century, the now obvious negative impacts on the city from car use have led to a new planning approach based on the functional redistribution of space and infrastructure. This has led to a more balanced use of mobility resources aimed at rebalancing the order of priorities among the various users of road space and also of public space to the benefit of sustainable and collective travel. This process of spatial redistribution is being complemented by other changes related to new technologies: electric mobility, sharing systems and autonomous driving. All together, these innovations are opening up new mobility options, such that the boundary between public and private transport is becoming much less pronounced. So what does the arrival of these new technologies mean for the city? Can we imagine that they will also lead to radical change in how we use means of transport?

A.B. Autonomous driving technology, although we think of it as applied to cars, can also be applied to buses. The two systems, the shared vehicle and the bus (i.e. a shared public transport system), are intertwined and the aim will be to achieve the advantages of both. For example, the autonomous car has the advantage of being unrestricted, with a user-defined route, while public transport has the advantage of a huge capacity that the car does not have, but is limited by having a predefined route. Trying to combine the advantages of both would open up interesting scenarios such as that represented by fleets of medium-sized vehicles (for about 10 people) capable of rearranging their own routes, where the route would be defined instantaneously by users as they get on board. This could be an example of how to merge fleet transport with public transport. There could be multiple scenarios which would open up a market for a wide range of services dedicated to different travel classes and comforts. For example, one user might want to reach his destination in total relaxation travelling in a small car in the comfort of an armchair, while another might want to adopt a cheaper solution and share a larger vehicle. In both cases, the technology will still be the same: that is, a vehicle that changes its route according to the customer's wishes. What changes radically is the context: there will no longer be public and private transport, but private transport will become fleet-based and public transport will come closer to the standards of private transport.

E.O. In rural parts of the territory—often perceived as residual and excluded from the most vibrant economic and social dynamics—not only important segments of the population, but also environmental and cultural resources with unexpressed or even endangered potential are concentrated. Can we imagine the extraordinary impact on these contexts brought about by new technologies that may introduce new available transport options and, as a result, changes in people's travel habits? Again at the territorial scale, how might new technologies provide an opportunity for more democratic and inclusive mobility that has less impact on the territory, environment, and economic resources?

A.B. This is an interesting discussion point. I was talking earlier about the possibility of not even owning a garage anymore. However, if this concept was pushed to its limit, it would open up an even more wide-ranging mobility scenario that could encourage people to move out of cities. At present, however, we are witnessing the opposite situation, people are moving to the city because they have the facilities and all the services they need within a confined area.

By contrast, with technology that allows for highly efficient mobility, we could benefit people who live outside the city by driving them to work in an automated vehicle, inside which they could already start working. This kind of mobility should be organised so that people can move outside the city limits, with important positive side effects, comparable to the digital divide I mentioned at the beginning. In this case, too, a third party is needed to govern the technology and to have a clear idea of the pros and cons of all possible scenarios. It would be a good idea to favour new companies that want to create fleets of vehicles and deploy them first in the peripheral areas and afterwards in the central areas of major urban centres. This would incentivise those who want to move automated mobility outwards from the city rather than those who want to deploy it within the city where business is greater.

E.O. Currently, the backbone of territorial transport on a national scale is based on the rail infrastructure. It is a heavy, land- and landscape-hungry system that offers movement channels limited by the rigid layout of the tracks. Electric propulsion, together with connectivity between vehicles and infrastructure, can decouple mobility services from rigid, predefined routes and ensure a greater coverage of the territory with benefits that can have positive knock-on effects in various sectors. In addition to this, assisted driving and autonomous solutions are making it possible to extend the safety standards guaranteed by constrained systems to road transport. Is this shift from fixed systems based on rigid infrastructure to distributed demand-adaptive systems possible, in order to export levels of service and coverage imaginable in urban areas to low-density rural areas ?

A.B. Automated driving technology shortens distances. To demonstrate this, I usually use this example: if I want to travel to Paris from Milan in a day to attend a meeting, my only option is to go by plane or train. With the rise of technology and the implementation of autonomous driving in vehicles, scenarios will change radically, offering additional possibilities for travel. Today, it would be unthinkable to drive your vehicle through the night and attend a meeting in Paris the next day. Once self-driving technology is in place, functional and well deployed, one can imagine getting into a vehicle, eating, sleeping, and the next day arriving at one's destination ready for the meeting. All this shortens distances because it opens up new scenarios that were previously unattainable. These profound changes will bring with them many dynamics that will impact a variety of contexts, not only safety and urban design. The entire transport industry will be affected: from air transport to rail transport to insurance companies. All these dynamics and situations that will be unleashed will have to be taken into account by leadership that knows the technology inside out and is able to understand its areas of impact and offer a future vision of it that is useful and usable by the entire community.

E.O. You have been dealing with automotive ethics for many years and with the fact that automatic vehicles will have to be equipped with decision-making autonomy in dangerous situations. In an interesting experiment by French engineer Matthieu Cherubini in 2013 entitled Ethical Autonomous Vehicles, three distinct ethical/behavioural scenarios are hypothesised for dangerous situations involving self-driving cars: the first scenario is that of human ethics, where the algorithm divides the damage caused by the accident among the people involved, protecting the weaker and more exposed people the most; the second scenario is that of protectionist ethics, where the vehicle attempts to protect itself and its driver at all costs; and the third scenario involves a profit ethic based on minimising costs incurred by the insurance company.

Do you therefore believe that it is necessary to legislate on the behavioural ethics of the cars of the future before they are put on the road, and with which behavioural approach? How do we integrate ethical issues, and which ethical frameworks should be integrated (the manufacturer's, the designer's, governmental etc.) into technological devices?

A.B. What we are talking about is the first example, in the history of mankind, of a system with autonomous decision-making capacities that will have an impact on people's lives. This is a key consideration in realising that those working in this field are pioneers. It is an extremely difficult problem to tackle because these are scenarios that have never before been brought to the attention of the community and therefore require rules that are well defined beforehand. The simplest rule that can be defined is the one concerning minimisation of damage if the vehicle is involved in an accident: let us suppose that the vehicle is moving in a traditional setting, an empty road, and at the last moment two road users, two motorcyclists, cross the road. The vehicle has to decide where to impact and the two motorcyclists look similar, but one is wearing a helmet and the other is not. The question for the vehicle at this point will be: which user should I crash into? The obvious answer would be to impact the motorcyclist with the helmet, who, thanks to the protection, will have fewer consequences and thus be more likely to survive the impact. However, this will generate undesirable behaviour caused by fear of collision among people who wear helmets as opposed to those who, in order to avoid this danger, decide not to use them. So even the simplest rule like this one—trying to create as little damage as possible—has hidden issues that need to be discussed. Vehicles are being designed that will find themselves in dangerous situations and must therefore also be programmed to handle those situations. Driving schools do not currently teach how to behave in the event of a possible crash or dangerous situation on the road. Self-driving vehicles, on the other hand, will have to be designed and defined clearly and comprehensively regarding all the dangerous situations they may encounter.

I want to emphasise that it is difficult to find an ethical rule on which everyone agrees. Moreover, I don't think vehicles equipped with advanced autonomous driving systems are always able to understand all the dense information in the environment and sufficient detail to make an informed decision. Returning to the previous discussion, in order choose between a motorcyclist wearing a helmet or one without, I have to assume that the vehicle is capable of understanding whether or not a person is wearing a helmet. To date, technology allows basic perceptual capabilities that enable the vehicle to tell if there is an obstacle and to interpret the direction of the road. However, it is not yet possible for these vehicles to deduce the age of a person or even to determine whether that person is wearing safety equipment. For now, it is impossible to perceive these details that are necessary to make an informed decision.

Furthermore, all these plans centring on vehicle ethics are based on a consideration that I think is wrong since they are utopian situations in which it is necessary to decide which user or obstacle to impact. Instead, I think there are many other alternative solutions that could be the best option. For example, I could swerve a little and avoid both individuals. Having said that, it is very difficult to read the environment and define which manoeuvres are possible and which is the optimal one to perform.

Personally, I think a convincing solution is one that more closely follows the way we normally drive, which is characterised by predictability. As long as the vehicle manages to be predictable, we avoid accidents. What we should teach in driving schools is to be predictable so that other road users understand our actions for the next 3–10 s. If we could be predictable, we would most likely see a decrease in road accidents. There is a hidden concept in predictability, namely that it changes according to driving culture. Predictability is the result of how people generally behave, so it is difficult to define, but I am pretty sure that if we can use it in our favour, we would have a useful parameter that would allow us to decrease accidents on the road.

E.O. Of course, predictability is also influenced by the number of self-driving vehicles that might be on the road at any given moment. The more autonomous vehicles on the roads, the more we will be able to predict their behaviour. The problem, if anything, will be caused by vehicles that do not have autonomous driving and therefore will not be predictable.

A.B. This is very true. In fact if we were able to switch overnight from the current situation to fully autonomous driving, we would have already solved most of the problems. Unfortunately, we will get there gradually, and therefore autonomous vehicles will have to drive side by side with manually driven vehicles, with all the unpredictability and incorrectness of manually driven vehicles. It is obvious that in the future, vehicles will have to be autonomous and manual driving will be prohibited because it is too dangerous. Driving is the most dangerous action we undertake on a daily basis and in the future it will have to be eliminated, allowing us to have fun behind the wheel only in purpose-built tracks and in total safety. Today, the vehicle is both a means of transport and an object of enjoyment, whereas these two aspects should be separated and the former should be guaranteed to be absolutely safe.

Conversation with Jurist Giovanni Maria Flick

Introduced by Dario Costi and Antonio D'Aloia, interviewed by Andrea Fanfoni
10 December 2020

DARIO COSTI Giovanni Maria Flick has written a very important book, "*Elogio della città?" (In Praise of the City?)* in which he talks about the city as a social organisation defined by the Constitution, an entity that is the place and instrument through which the individual can realise his or her personality. In this constitutional definition relies the importance of the city environment and perhaps reasoning about the right to the city is an issue that has a constitutional foundation. For us, this scientific and cultural stance is very important because the constitutional law, when addresses the city, is an issue that affects everyone and that all citizens. I asked Professor Antonio D'aloia, who is a member of the research laboratory, to bring a greeting and introduce the professor.

ANTONIO D'ALOIA This is not the first time that we had the chance to have a conversation with Professor Flick about contemporary issues of our society, addressing them in a way that is never only juridical. Moreover, Professor Flick comes from an extraordinary experience in his field because he embodies all the facets of legal science. He has been a magistrate, lawyer, academic professor but also a statesman with important institutional positions. I would like to mention his work as Minister of Justice, as Constitutional Judge and as President of the Constitutional Court, which is the highest court of our Constitution, the voice of the constitution as someone said. It is important to underline Professor Flick's work in these years: the construction of a civil lexicon, analyzing the dimensions of law comparing them with current problems and progresses of our societies. The book "*Elogio della città?" (In Praise of the city?)* is the latest, only in order of time, of writings that speak of the city, the constitution, dignity, justice, heritage, passing through the essential categories of our constitutional identity and the problems of our community life. Today we are going to talk about the city and we are going to do it through this new and interesting form of a cross-disciplinary conversation. The city is a place that refracts the vitality of constitutional values. In his book, Professor Flick says that the city is one of the modes in which the society is organized, one of the modes that Article 2 of our Constitutional Law includes between the premises for the protection of human rights, not just as an individual, but also as part of social formations. The city is space, but it is also space where time has its own history. Space and time are the two main coordinates of human life and therefore of law. The city becomes a place for reconstructing an identity, a memory, a past, and a history are oriented towards the great future challenges. Today, the city is the place where verify the great questions of our times: sustainability, the challenge of inequalities, the issues related to creating smart inclusive cities, amplifying the constitutional value of solidarity. In this perspective, the city is not only a mere part of our administrative units, from metropolitan cities to small municipalities, but it is the place where the constitutional principles, in particular the principles of equality and dignity, are tested. The city in its wider meaning, from megacities to informational cities, will always have to take into account the great axiological and value trajectories that are fundamental to not lose our identity.

ANDREA FANFONI I would start with three questions, on three concepts, that we believe are key elements of reasoning for our research on the city. The first question leads us to the word of technique. Technique was born as an expression of the human spirit and according to the Treccani encyclopedia it is the practical ability to operate in order to achieve a given end. We think that the analytical and dialectical construction of a "culture of technique" is a central issue of contemporary thought, and not only on the city. At every industrial revolution, critics have questioned the risks of a possible subordination of human beings to technical instruments. Before us, thinkers such as Walter Benjamin, artists such as Paul Klee, theologians such as Romano Guardini and architects such as Mies van der Rohe have questioned the risks of the rising modernity. Today, the critique of what Arjun Appadurai has called an "increasingly and pervasive dependence of the world on digital networks and mobile technologies (as well as the hidden structures on which those systems stand)" becomes more urgent than ever. Thus, the development of ICTs (Information and Communication Technologies) has increased the level of control made by our technological devices, creating autonomous mechanisms. How can we think today of a more thoughtful "culture of technology", which the people are at the center, developing the necessary skills and tools to start a re-appropriation of this knowledge by individuals and communities?

GIOVANNI MARIA FLICK I would like to start from Dante: "*de' remi facemmo ali al folle volo*"[10] that guided Ulysses to cross the Pillars of Hercules to reach a dimension that responded to the famous motto "*fatti non foste a viver come bruti, ma per seguir virtute e canoscenza*".[11] This is a necessary premise at a time in which the evolution of our way of life and our thinking is oriented much more towards knowledge than towards virtue. Then there is a crucial question that arises from your definition of technique: the ability to work towards an end. But what end? What perspective should we think about?

I think it is useful to return to the past to be more concerned with the future. In Genesis, there is a beautiful page in which Roberto Calasso[12] describes the Garden of Eden: "*Yahweh Elohim planted the Garden of Eden and placed there the man he shaped. In this place he made trees and plants growing and man was placed to take care of it. Of all the plants Iahweh Elohim pointed to only two: the tree of life, in the center of the garden, and the tree of the knowledge of good and evil, which is not told exactly where it was. Yahweh Elohim prohibited to eat only the fruit of the tree of knowledge.*" "*To those who ate its fruit, of the two trees that Yahweh Elohim indicated them, they would assimilate two qualities of Yahweh Elohim: knowledge and eternity. If the man and woman had went to the center of the garden, looked at the tree of life and eaten its fruit, nothing would have changed. Life would have continued exactly as before, but the man and woman would have access, once and for all, to an*

[10] "We of the oars made wings for our mad flight".

[11] "*you were not made to live as brutes, but to follow virtue and knowledge*" in Dante ALIGHIERI, *La divina commedia* - inferno", XXVI.

[12] Roberto CALASSO, *Il libro di tutti i libri*, pagg. 349–350, Adelphi Edizioni, Milano, 2019.

everlasting life. While concerning the quality of the everlasting life Yahweh Elohim did not raise problems, he raised them on the knowledge of good and evil: "Attention was focused not on life but knowledge". The snake attracted the woman, which convinced the man who let himself be convinced, they ate the fruit of that tree, and "the consequences were immediate and perceptible. The man, said Yahweh Elohim, addressing himself to whom we do not know, had become like one of us. It was immediately necessary to prevent him from stretching out his hand, taking even the tree of life fruit, eat of it and live forever". It was at this moment that man became mortal. He preferred knowledge to immortality even without any prohibition to take from the tree of life because Yahweh Elohim would have never tolerated that man became like him.

This argument is joined to another very poetic and very clear statement that takes up the theme of the city of Pietro Citati[13]: "*In the New Jerusalem, we no longer find the tree of the knowledge of good and evil, from which Adam and Eve had eaten the fruit". "The tree of knowledge had divided the universe according to the opposite forms of good and evil: it had established in the unitary world of the origins the separation and the antithesis: on one side good and on the other evil, on one side the sacred and on the other the profane; and then the pure and the impure, virtue and sin, permission and prohibition, law and violation, life and death*". There remains only the tree of life at the end in the New Jerusalem. I believe that these two points are quite important when we are going through a crisis that leads us, through an evolution of science and technology, to a radical change in our way of living together. The transition from the analogical society to the digital society, that is, the exaltation of technical knowledge through the modifications of not only behaviors, but of the fundamental structures of our society. The analogue society is destined to die out; it is the older people heritage. What is certain is that the transition from analogue to digital society has overwhelmed the representation. Digital society is characterized by disintermediation and direct democracy (it remains to be understood direct by whom and direct where). We have yet to understand, decide or clear our purpose and direction.

A.F. Professor, even the word smart is at the core of our cultural debate.

Its meaning opens to interpretations that span from clever to vivid, but also quick and fast. In urban studies has been coined the Smart City neologism to refer this concept to cities. In smart models, we see several limitations that derive from the reduction of the land governance to its technical dimension, from the externalization and privatization of public utilities, from the deepening of the digital divide. How can we decline the intelligence for the city, developing citizenship rights, and reducing inequalities?

G.M.F. Let me discuss digital cities, which are very interesting for me, also because of the dangers related to digital society. A society in which, for instance, institutional intermediations disappear with the delegitimization of Parliaments by digital platforms and where all the features slowly conquered during the democratic process are exposed to a lack of consideration. This opens several legal problems on how to regulate this new digital society in which human pride has led information, as a tool, to become an autonomous form of science. Information technology opens the way to several issues: robotics, the internet of things, machine learning. Just these days we are talking about the fact that the weight of human products exceeds the weight of existing situations in the world. Today the product of man exceeds the product of nature.

I speak with all this consideration about the digital society because it is useful to remember that the digitalization of society is one of the first inequalities not introduced, but exposed and highlighted by the pandemic.

There is a substantial difference between the elderly and all young people, that is the gap between those who are growing up within a digital society and forming themselves around it and those who have remained anchored to the analog society and struggles with the new

[13] Pietro Citati, *L'armonia del mondo*, Adelphi Edizioni, Milano, 2015.

devices. This disparity leads to a huge problem of inequality, worsened by the pandemic, which has replaced human relationships with digital connection since when we theorized that the only form of reaction and defense is avoiding physical contact. The physical connection was the main reason for cities' inception, although with other reasons such as defending against fear.

From this point of view, cities have betrayed their historical duty. Cities were born to protect us from the fear of what was happening outside but we did not realize that slowly the fear entered inside the city through the pandemic. The external walls fell into disuse, those that were supposed to defend men from external adversities, from the climate, from the beasts. Today the real problem is that fear is inside the city. The violent city, the unfair city, up to the Berlin Wall paradox. A wall that divided in half streets, facades, and windows.

This point is crucial because the reference to the pandemic and the need to avoid contact was the premise for greater effectiveness, actual or not, of digital connection than human connection. This issue, I don't know if because it is conditioned by my analog origin, is of great concern to me.

There was also a shift from the alternative of doubt to the alternative of certainty. Either yes or no. Either black or white. There is no longer that situation of ambiguity and uncertainty that, for example, in the field of justice is a form of salvation.

Let's think about one of the most interesting concepts in law, I speak as a criminal defense lawyer, that is the reasonable doubt for the sentence or the reasonable duration of the trial, fundamental values embedded in the Constitutional Law. What is reasonable? Where does the reasonable begin and end and where does the unreasonable begin? It is clear that the problem is the importance of learning and adapting, to cope with the new reality. And we have only two extreme ways of dealing with it: refusing it or accepting it as it is and exalting it.

In the field of law, we have proved this by giving up the principle of legal certainty through the law. For instance, I personally started from a search for certainty through law as a matrix of certainty; I come to understand that the law has nothing certain about it, indeed I have discovered that true maturity is that which moves away from dogma and accepts the idea of constructive doubt. Then I believe that the third way between accepting or rejecting is to try to adapt. The real strength is not the characteristic of the brightest and most capable races, but it is characteristic of the most adaptable races. The figure of the future is the mixing of mestizaje.

I make another important point to me. We have experienced every night on television, through the debates of virologists and doctors, a series of arguments that have led us to a complete misunderstanding. Science, which was supposed to be the starting point for politics to take decisions, has emerged its conflicts in public. It has been said that science describes, examines, studies what is there. Politics describes, examines, and tries to act the choices to build a different reality. Science is knowledge, politics is decisions. Let us think about the confusion that we experienced last winter when it was no longer clear whether the decisions were made by politics or by science. From this emerged the danger that technology, even worse than science, would prevail.

A.F. As a dialectical and cultural counterpoint to the smart concept, often simplified and trivialized, we are thinking about the idea of Wise City. What alternative approach to technology can develop a city which is not only fast, speedy, and connected but also wise, fair, and inclusive?

G.M.F. What is the pathway that we imagine for cities? The city, as I have already mentioned, grows from fear and violence. Taking up the Bible, the first city, Enoch, was named as the son of Cain and was the emblem of violence. The second city is Babel, the city of arrogance, which ends with punishment and the mass dispersion of the people to the world, representing an advantage for the development of new ideas and pluralism. The third city is Sodom, which ends badly, and is the emblem of the exploitation of the foreigner, the guest, the different; then for sodomy, today for the harvesting of tomatoes, but the concept remains unchanged. The fourth city is the city of profit with the great ports overlooking the Mediterranean Sea of Tyre and Sidon, for example. The fifth city, the one of which Pietro Citati talks about, is Jerusalem, which at the same time combines in its ambiguity the characteristic of being the Sponsa Dei on one side and on the other the great whore. Jerusalem is also the city in which, when it becomes the heavenly city, the problem of the knowledge of good and evil disappears. In this path, I believe that the city has betrayed man in a fundamental way: it was born to fight fear, but it has become its elective seat.

The city is like a company, which produces services, sewers but also cultural institutions and hospitals sometimes forgetting those in the countryside. After my experience at the Constitutional Court, for about a year I was called to lend a hand to prevent the collapse of San Raffaele Hospital. This large scientific-medical facility in Milan was created with a lot of resources and a lot of wastefulness. The sensation I had while working on that problem was that of a healthcare system that was isolated above all in the main urban areas, forgetting about country healthcare, everyday healthcare, and home healthcare. A problem that has presented much more dramatically, when we experienced this pandemic period. If we look back, history gives us the lessons we need.

The dream of the Smart City, which through technological solutions finds well-being and happiness I worry is a disappointment.

Also because a Smart City that stops at this point is a Smart City in which the problems of the fair city, of cohesion, of coexistence, of sharing, find a limited solution. So, if I had to give a press subtitle, I would say that the Smart city is the city of illusion. Technology is necessary to achieve conditions of livability and urban mobility. Let's think of what happened two years ago when the whole Liguria Region went into a deep state of crisis due to the fall of the Morandi Bridge, understanding the real meaning of a divided city. I don't think we can trust only the in Smart City. Then there is the step of the Wise City: that is a city that achieves balance. We have to be careful, however, because there is a threat: the city of balance, which requires rules to guarantee it, could only end up becoming the city of institutions, of bureaucracy, of the relationship between local and central power. Here, too, we have a formidable lesson from the pandemic. The relationship between the State and the regions and the chaos created in the clash between the central and local dimensions has resulted in great confusion in the regulations, causing a conflictual relationship in the resources management. The solutions found by the Central State have been ineffective, such as the constitutional reform of our country in 2001 in which the Central State is responsible for the general principles and the Regions for their applications. These days we have realized how much this is a problem. The Constitutional reform was hurried and did not look at the Constitution itself but echoed political needs of not displease the current political powers hoping that they would be satisfied. I think the Wise City is still a utopia, and we should still work hard to get there.

A.F. In a recent interview to the HuffPost the dated 24 April 2020 you consider the Constitution as a survival manual inspired by solidarity values against the Covid-19 effects to not leave our cities to five main inequalities which affect jews, migrants, women, and also elderly and inmates after Covid-19. Even Pope Francis, in his first public speech after the easing of lockdown, talks about solidarity. He says "To get out better of this crisis we need to work together, awakening solidarity, which means much more than a few sporadic acts of generosity and requires a new mindset with a community thinking". Thus, I would ask you if, during these months, something has been made in this direction and which are the current priorities.

G.M.F. In this sense, I was very touched by a statement of Rita Levi Montalcini, who is the emblem of the differences that the pandemic has worsened. She was a woman, and we know how the pandemic has worsened the female condition because it implements in coercive terms the saying Kinder, Küche, Kirche (children, kitchen, church)[14]; this leads to a negative step back to the development of women conditions. She was a migrant, she had to leave Italy for Belgium to avoid persecutions, and she was a jew, belonging to a "different race" which stressed her condition of diversity.

Within a society that builds a system of connections rather than relationships, we have two other categories of unequal people who seem to be significantly impaired by the pandemic: inmates in prison, condemned to forced cohabitation, and the elderly, who dies in solitude not even finding a proper place. In this situation, the plea of Rita Levi Montalcini seems very important to me. When she was awarded with the Nobel Prize, she gave an outstanding speech in which she wondered why there was only a charter of rights and not a charter of duties.

I would cite another example that seems quite pertinent to me, thinking about the pandemic we are going through and particularly about the right to refuse drugs. When our Constitution was written, the memory of the trials of Nazi doctors and their ignoble experiments was vivid and therefore was crucial to state the self-determination faculty in rejecting drugs. Later it raises the understanding, and the Constitution pointed it, that health is a fundamental right not just for the individual but that also for the community. Therefore, the right is coupled with the duty. Let us think about vaccination issues, no vax, and the debate about this type of progress. The radicalization between the different standings of "yes" and "no", between obligation and refuse, is a symptom of the problems we are facing in a society that accepts only these two standings without evaluating intermediate positions.

I would conclude with a concerned thought about lockdown. This refusal of contacting others, if not mediated by digital connection, ends up to a state of worrying isolation for the man, in which he loses his natural ability to interact with their equals. Contact is limited due to the way it takes place and to the contents that it express. For example, it is very challenging to communicate a poem using digital technologies. It is much easier to express a logarithm. I believe that these elements will be very important to deal with not only the pandemic but also our after-pandemic society.

A.F. In your book *Elogio della Città? (In praise of the city)* you talk about the need of more urban culture, going beyond a detached urban planning approach through an organic strategy of retrofit projects. The city dies if it is not wise and inclusive and if it forgets the memory of its own past. You said that we are losing our cities. Is there still time for this not to happen?

G.M.F. There is a third city typology, very important, in which the joining link is the reconciliation between man and the environment. On one side, between man and nature, on the other the understanding of the fundamental principle of solidarity. We should remember that there is no equality without solidarity and solidarity is, citing the Constitutional Law, what permits to reach a balance of fair social dignity through the removal of the obstacles generated by the abuse of power and discrimination.

I am admired by the effort of Pope Francis to propose us again the Canticle of the Sun, remembering us the forgotten values on the importance of nature and the environment. We celebrated the 2015 Expo using as the main symbol not one of our thousand-year-old trees, which abound in Italy, but a tree made of nickel, steel, and LEDs.

I would have preferred that the Expo be represented by a thousand-year-old olive tree that we have in Puglia. The third hypothesis is therefore the Green City, which we have tried to do in some small experience. Urban green is not only the treehouse or a condominium with greenery on every balcony. The urban greenery is also the real gardens, become during centuries the heritage of the whole community. A beginning of rethinking about the public space. Here there is a great task for architects, graduating the

[14] It is a German slogan used to indicate the three values that a righteous woman in western society should respect, namely family, household chores and Christian values. (Source: Wikipedia).

limit between public and private space, balancing the communal use of space with the space in which each one goes home. It is very interesting to discuss the value of urban green spaces.

A.F. Always reading your book *Elogio della Città? (In praise of the city)* we feel the need to find again an ethical conception of the city, in which the city is consideres as a "common". In your opinion, which are the main concepts on which we can begin a reflection about the relationships between ethics and city? How can architecture ease "the participatory governance of commons"?

G.M.F. In addition to the city as a common good, I would also include the forest as a common good. The mountain communities, which had the right to clear their land, to gathering firewood, to shared use beyond the formal label of public or private property, anticipated the discourse of common good, that is, those goods that must be guaranteed to all, beyond property and the civil law schemes born in the century of property. In the declaration of human rights, *Liberté, Égalité, Fraternité,* fraternity is lost along the way and becomes property. From the century of property, we have passed to the century of information without taking into account that link which is solidarity and which was fundamental to continue living together. This is my idea of the relationship between the city and the forest and the idea I have of the future if you will not deal with the problem better than we dealt with it.

A.F. Sociologist Evgeny Morozov considers the smart models narrative simplify urban issues by reducing them to their technical dimension. This process would be necessary for private IC companies to present their offer as essential. What do you think about the relationship between ICT technologies and the city? Can urban planning models based on technological development be a model for our future cities?

G.M.F. One of the tragedies we are experiencing with globalization, with the cult of efficiency, and with digitalization is that we live mainly in the present. With virtual digital relationships, we have abolished the problem of the past and the problem of the future. The right to the memory of the past and the right to hope for the future are the conditions that we are losing in the perspective of presentism. A condition in which the main factor is profit and in which we have the illusion of not becoming a sort of machine necessary to make work the whole Big Data system. A system where we are, consciously or not, factors manipulated by someone else to return us fee forecasts.

A.F. The Italian urban context, and specifically the one of Emilia, is mainly represented by a series of small and medium-sized cities widely connected with the countryside. Each of these cities has a deep historical, cultural, landscape and enogastronomical identity. Paraphrasing the architectural historian Joseph Rykwert, all these aspects linked to geography, climate and culture explain the way we live in our cities, which is the result of thousands of years of settlings. Every urban form has its own uniqueness, its own character, a specific reason that finds in specific buildings and special structures the places recognized by its own community.

Today, faced with interrupted, incomplete and in many ways wrong places, urban centers are to be rethought starting from the revival of an idea of the city, for a valorization of their identities and a unique urban image valuable for the promotion of these contexts. What model of city, intended as a community, is therefore necessary to imagine for a revival of excellence and territorial promotion?

G.M.F. Answering your question, I would like to return to the relationship between the city and the forest. The forest is a symbol of the environment. The environment is formed by biodiversity and by unity in diversity because the forest is unique, but within the forest there is a surprising variety of life. The forest is the return to nature and we have to find a balance between excessive care—what I call Taliban environmentalism where nothing is touched—and uncontrolled destruction. The forest must be cultivated. I am always reminded of Dino Buzzati's beautiful novel entitled *Il Segreto del Bosco Vecchio ("The* *secret of the Old Woods*"). Buzzati says that Il Morro (one of the main characters, translator's note) had a lot of properties, but when he died it turned out that there was a part of the land where trees were not cut down, and continued to live and be cared for. The heir, Colonel Procolo, when he arrived, said that everything had to be cut down. But then he understood, through a dialogue with the trees and the wind, that it would be more convenient for him to use the forest reasonably.

I wrote a book called *Elogio della Foresta. Dalla selva oscura alla tutela costituzionale (In Praise of the Forest. From the Dark Forest to Constitutional Protection)*[15] (after "*Elogio della città?*") precisely from this perspective. Instead of delirious cities, the alternative of nature is raising its appeal once again, and I believe that the trend of urbanization is now reversing. We know that by 2050 80% of the population will live in the city and the remaining 20% in the countryside and forests. I will not be there for reasons of age, but if I was here I would be in that 20%. If only to recover the light and sound pollution that has become constant for us.

I wrote this last essay precisely because it seemed fair to recognize among the many negative things that these governments have done at least one positive fact: the history of forest regulation passes through several states, the first of which is the hydrogeological protective restriction. The realization came when we understood that deforestation leads to the loss of soil because of the rain, leading to the consequences we are seeing. Disasters such as floods and landslides. For every ski slope that we prepare in the mountains by deforesting we put a mark on the floods that will follow it. The Serpieri Law of 1923, with very strict rules to avoid hydrogeological instability, tried to avoid these catastrophic events. We then find the recognition of the link between the value of cultural and artistic heritage and the value of environment in the Bottai Law of 1939. Then the Constitutional Law, in which article 9, placed among the fundamental principles, says that the State must protect both the past, intended as the national historical and artistic heritage, and the landscape, given future improvements. What the Constitution calls landscape is all the environment. After that, the recovery of a reasonable productive dimension and the premises for a balance between State and Regions in the management of the forest heritage vanished with the constitutional reform of 2001. Finally, the Constitutional Court recognized and emphasized the value of the landscape and the forest and came in 2018 a Forestry Law (D.Lgs 34/2018—TUFF) in which the lawmaker has tried to balance the profit and the environmental needs. These two dimensions in the city have been completely lost and remain only in the countryside.

A.F. Thank you, Professor, for this valuable contribution. I turn the floor over to the professors for their final remarks.

A.D. It was a beautiful reflection on the city, not only as a physical place, as a space in which general issues, conflicts, and problems of contemporary society gather, but also of the idea of the city as a community that must find itself in the real and dynamic definition of the identitarian principles of our constitutional history.

G.M.F. The place where the personality of the individual takes place, in a social formation in which the Republic recognizes and guarantees the inviolable rights and the inderogable duties of political, economic and social solidarity (art. 9 of the Constitution)!

A.D. Exactly! The aspects you raised certainly gave proper answers to the doubts that were also emerging from the questions posed by the Professor Costi research group. Questions about the future of the city and about the city as a microcosm of what we want to be soon. I have appreciated very much this discussion that takes place on a temporal range that is continuously shifted between past and future, which is the dimension of law and the Constitution. This is because they live in the past maintaining the ability to design the future and to

[15] Giovanni Maria Flick, Maurizio Flick, *Elogio della foresta. Dalla selva oscura alla tutela costituzionale*, Il Mulino, Bologna, 2020.

respond to the future problems that will arise in the history of humanity and social coexistence.

On the concept of heritage, I would augment that heritage implies transmission. Transmission, in turn, presumes conservation, that is, a way of using things that consent to preserve them and make them usable in the future. I believe that this link between city, common good, heritage is really important to define this framework of values.

D.C. I am very pleased with the outcomes of this conversation and with the contribution to our research.

G.M.F. I would like to thank you and I hope that sooner or later we will have the opportunity to meet and discuss in person these issues that are fundamental to our existence. I am thinking, for example, about the hybidization between the market and the environment in the challenge to reduce CO_2 emissions and avoid climate change …

Conversation with the Philosopher Sébastien Marot

Introduced by Andrea Borsari, interviewed by Federico Diodato
12 January 2021
[Text not edited by the interviewee]

ANDREA BORSARI I would like to include, within this chosen panoply of disciplines, the philosophical point of view with reference to the dialogue we had in the video interview with Sébastien Marot, a philosopher who works at the École d'Architecture de la Ville et des territoires Paris- Est, who has taught in various European and American universities, a scholar who introduces that dimension of reflexivity typical of the philosophical attitude of questioning the very procedures of knowledge adopted with respect to a thematic field, in this case regarding the problem of technology.

An enormous problem that can be variously discussed and tackled, but which Marot appropriately proposes to focus on as a problem that should not be taken en bloc, accepted, rejected, demonised or apologised for.

The specific point on which his reflection is based is that which reflects on the mismatch—using a terminology that is not his own but which is of interest to us because it seems to me that globally with respect to the digital world it has become a problem that concerns everyone—between what we could call, following Georg Simmel, objectified culture and subjectified culture, that is, between the amount of knowledge that is incorporated in technical objects and in particular in digital objects and the amount of knowledge possessed by those who interact with these objects, that is, human beings in general.

The growing gap between these two dimensions is something that was noticeable even at the beginning of the last century, but which has become, thanks to the latest technological revolution, obviously unbridgeable and growing.

What reaction can we adopt, still in that reflexive dimension typical of this look from Marot's perspective, but also from a more general point of view, adopting that capacity to see the heterogenesis of ends that—as a philosopher he quotes, Pierre Caye in his book *Durer*, suggests—produces an abstraction in which the digital puts the city at its service, so that this heterogenesis of ends pushes us to a reaction.

This is a cognitive reaction but also a practical one; the cognitive one we could define on the axis between the global and the local, the practical one a capacity to act locally, which in Marot's case, which is not an eccentric case, can take concrete form, for example, in the reappraisal of those sustainability practices known as permaculture, i.e. involving a capacity for self-sufficiency and maximum reduction in consumption for the feeding of a given community.

These global and local axes also contain another theoretical move, which is to recover Gilbert Simondon's contribution, which frames technology as a sort of further kingdom, that is, as something that functions, with respect to the animal, vegetable and mineral kingdoms, with autonomous laws and that produces within itself objects and successive generations. This perspective insists on saying that rather than the city, our point of interest becomes a world composed of other worlds in which technological resources tend to performatively construct an autonomous sphere around themselves.

This is the reasoning that Marot proposes to us, which seems very interesting from my point of view thanks to the capacity that he identifies to construct a legibility of the local, therefore to insist a lot on the necessity to disaggregate this macro-object in which we would be thrown as in a destiny according to the prevailing perspective of the reading on the problems of technique.

Instead, to disaggregate it, to know it locally, which however maintains the connection with that world that is the frame and the overall connection.

From this also derives the ability to question ourselves on which worlds we would like to build with technologies, through a dimension of reflexivity applied to this sphere, trying to put ourselves also in the local attitude of repairing and recycling and the kind of practical attitudes that derive from it.

This is a bit of the general outline of the discourse, it seems to me that it has made a specific contribution to our reflection, then we will see how to develop it in the future, but it seems to be a plan, even before a contribution of specific contents, ineliminable from our research.

I give the floor to Federico Diodato, graduate student of our group of universities in Emilia Romagna in co-tutorship with Eav&t Paris-Est.

FEDERICO DIODATO The first question is about technics. Technics arose as an expression of the human spirit, and according to the Italian encyclopaedia Treccani it is the practical ability to work to achieve a given end. We think that the analytical and dialectical construction of a "culture of technics" is a central issue of contemporary thought.

At every industrial revolution critical thinking has questioned the risks of a possible subordination of human beings to technical tools. Before us, intellectuals such as Walter Benjamin, artists such as Paul Klee, theologians such as Romano Guardini and architects such as Mies van der Rohe have wondered about the risks of the assertion of modernity. Today, the immanent criticism of what Arjun Appadurai has called an increasingly widespread dependence of the world in which we live on digital networks and mobile technologies is becoming increasingly urgent.

How can we now think of a more thoughtful "culture of technics" that focuses on people, developing the necessary skills and tools to allow a reappropriation of knowledge by individuals and communities?

SÉBASTIEN MAROT This is a very broad question. Since you're addressing it by introducing several references, I'd simply like to remind you of others.

One of the authors that I think should be reread carefully is André Leroi-Gourhan, who has made a technical history of humanity and the environment that goes far back in time. At first he was a scholar of prehistory, but he was above all an extraordinary historian of technics and what Alfred Lotka later called exosomatic organs.

This is a fundamental distinction:

In the first place we have endosomatic organs that are the object of technical development, there are techniques of the body that consist in mastering gestures. The history of technics is the history of the development of exosomatic organs, which we call instruments, organs that can take the form of a milieu. In this sense, other relevant references that come to mind and that we must draw from are those thinkers who have linked the history of environments to the history of the development of techniques. One of these is Lewis Mumford, who strongly poses this problem in his two brilliant books Technics and Civilization and The Culture of Cities, published in the 1930s.

To Mumford I would add Sigfried Giedion, who a few years later, in his famous text Mechanisation Takes Command, examines the importance—or rather the invasion—that technical development has assumed in the formation of the environments around us. And to these references I'd like to add yet another, very well-known in Italian academic circles: the philosopher Gilbert Simondon.

Simondon is part of a long series of reflections on these issues, that have their roots in André

Leroi-Gourhan, but which they developed starting from the Encyclopedie française. These reflections culminated in the 1950s in an extremely powerful and original reflection called Du mode d'existence des objets techniques.

He would continue to reflect on these issues, in this text and many others, where technical objects are divided into three levels: Technical individuals, technical elements and technical ensembles. Technical individuals are what Leroi-Gourhan and Lotka call exosomatic organs, the objects we commonly work with, i.e. a computer, a house, a building, a car, etc.

Technical elements are technical objects that are included in the composition of technical individuals, which are necessary for their constitution, and which we know less and less, obsessed with technical individuals. Finally, technical individuals never exist independently of the technical ensembles they're a part of. One of Simondon's fascinating theses is arguing that technical objects form a kingdom that is particular and distant from the modes of operation of the mineral, plant or animal kingdoms. In these realms, technical individuals give rise to a new generation of individuals in different ways. In his mind, at a given moment in the world of technical objects, a technical ensemble generates a new technical element, which will give rise to a transformation, or a generation of new technical individuals. This makes the world of technology fairly complex to understand. According to him, there's a "law of relaxation": for example, a technical ensemble corresponding to the agricultural society typical of the eighteenth century, allows for the birth of a new element, which for example is the steam engine. The steam engine, incorporated into technical individuals that already exist—mills, wells, wagons—modifies these individuals that thus form a new ensemble, the thermodynamic ensemble. A century later, the thermodynamic ensemble generates a new element, the electric motor, which will modify the technical individuals and therefore give rise to an electrotechnical ensemble. Later, this ensemble will replace the thermotechnical ensemble, which will give life to electronic magnets, that in turn will produce new technical ensembles. The question of the Smart City today has to do with the establishment or the project of a new ensemble.

F. D. With regard to this term smart, it's a term that today is at the centre of a cultural debate. The term embraces different meanings that vary from intelligent to lively, sharp and rapid. In the urban sphere, the neologism Smart City was coined and applied to the city. In smart models we see a series of limitations that derive from a reduction of territorial governance to its technical dimension, from the outsourcing and privatisation of public services and the widening of the digital divide. How can intelligence for the city be employed to develop the rights of citizenship and reduce inequalities? As a dialectical counterpoint and cultural counterweight to the often simplified and trivialised concept of smart we are contemplating the idea of Wise City. What alternative approach to technology can a city develop that is not only fast and connected but also wise, fair and more inclusive?

S. M. What we call smart is the development and colonisation of our milieux by a series of communication and transmission technologies. It's a question of connecting the distinction that you make between smart and wise: as you well know, smart means clever, intelligent, fast. It evokes technical solutions that are scalable, that work really well in any situation. They are technical routines that have a certain level of effectiveness in almost all situations, so they are often unwieldy.

The word wise, means a form of intelligence that also involves prudence, equilibrium, a balance of things. Wise is always a consideration that is fine and careful, it refers to the area of attention, the situation that we have to face. There's something very casuistic about wisdom.

What worries me is the danger that we'll rush into scalable solutions that can be bought, and that can work together with a fairly rapid adaptation. And therefore forget, aside from the effectiveness of the instruments that allow it to work, the very reason for the milieux that are transformed by these solutions.

F. D. The precursor of cybernetics, the mathematician Norbert Wiener, back in 1948 warned about the power of technological development: if technology fulfils our wishes to the letter, asking it to give us benefits, are we sure that we're asking for what we really want and not what we think we want? Referring to smart planning models, doesn't it seem to you that there's a risk that the actual impact on the region is different from what we'd hoped?

S. M. Yes, smart is a logic of supply for a number of effective technological resources that we should choose or that we couldn't do without, because others don't. It's the race to acquire the latest technologies. Obviously there's a risk in focusing the debate only on the technologies or the means of acquiring the most advanced technologies on the market, rather than asking the question of which worlds we should build with them. This is a very significant problem.

Simondon evokes it strongly in his work, even though he was a lover of technical objects, that he knew very well.

In his philosophy lessons, when he was a high school teacher, he had his student philosophers build televisions and radios, because in his opinion, it was very important for intellectuals and philosophers to know what they were dealing with, to know the technical objects, or electronic objects, so that they could talk about them.

It's therefore very important that intellectuals and politicians be careful when assuming technical objects as objects of knowledge. Simondon was also aware that these technical objects had a tendency to evolve towards the incomprehensible, and so he devoted an extremely important reflection to the question of access to technical objects. He stressed the fact that open technologies are technologies that allow both intelligence and hands to penetrate their internal constitution.

He also believed something that I think is important: technical objects must be known and loved, but for this to happen they must be lovable.

F. D. The technical object has always had the role of mediator between nature and the human being, however several thinkers and historians of technics have emphasised how in recent decades there has been a change in this relationship and that the technical milieu has imposed itself on the natural milieu. In the book Le Système technicien of 1977, Jacques Ellul argues that the technical milieu acts directly on humanity by asking people to adapt in ways that are comparable to those originally demanded by the natural environment. In this context that we could define as technical determinism, what do you think is our degree of "freedom"?

S. M. As a community and as individuals, we aren't always free to use technologies that we don't master or understand. But in our societies, recognising and living with this technical ignorance is a necessity that we can't live without. It seems to me that a considerable effort must be made to enable the processes that regulate the worlds we live in legible, to avoid becoming illiterate. This implies that we strive to acquire a certain understanding of the technologies that surround us, but also that we collectively control these technologies so that they do not become illegible in their operation and for the purposes they can serve.

This was the subject of an extremely important reflection that has been ignored even in the architectural environment it came from: that of Colin Moorcraft, a 25 years old architect of the 70s who describes what according to him should have been a post-industrial technology.

By post-industrial technology he means a technology that is used to build resilient worlds, highlighting three principles that should be present: the first principle is the principle of "cooperation", which indicates a post-industrial technical system in which each element performs different functions and each function is performed by several elements. After 50 years of dominion of functionalism, Colin Moorcraft proposes an alter-functionalism in contrast with historical functionalism, whose model was the machine where each element performed a single function but superbly, in an ultra-efficient manner. He replaces it with a system that is linked to the ecosystem and no longer to the machine, and for which resilience more than efficiency is a fundamental virtue.

The second principle is what he calls "integrated", a circular system where externalities are reintegrated as inputs and don't create a chain of negative or positive externalities. This is the idea of recycling, or more precisely of reuse. The last principle is the "flexibility" of the technical systems that can be repaired by their users. The concept of "user repairable" is more or less the same as the idea of intermediate technology, or alternative technology, or radical technology, which was so fertile in the 1970s.

F. D. Regarding intermediate technology, a concept of technology with a human face, which focuses both on human needs and limits, as well as the appropriate use of technology.

In the recent exhibition you curated at the Triennale of architecture in Lisbon, Taking the Country's Side, you refer to a particular form of intermediate technology: permaculture.

What do you think are the lessons of permaculture that we can apply to spatial planning?

S. M. Permaculture is an almost literal implementation of Colin Moorcraft's three principles that I just enumerated, when applied to the administration and the rational management of local food production. Permaculture is a syntagma that encompasses permanent agriculture. The idea is to design sites where each element satisfies each function and each function is satisfied by several elements. In these sites the inputs within the system are in a cycle and everything is an open, repairable technological system that can be directly administered by its users, minimising dependence on large technical systems. This seems fundamental to me because, while questioning the principles that have led to the industrialisation of agriculture, permaculture is at the forefront of a whole series of other movements, such as agroecology and organic farming.

In the field of architecture and urban planning, which in my opinion is not so advanced in the criticism of previous urban planning systems, permaculture is interesting not only for what it does directly, but also for all the extrapolations that can be made from its principles. This was the meaning of the exhibition Agriculture and architecture, where we reflected on what we can draw from permaculture today. To sum it up, I'll draw on the famous triad of Vitruvius and Alberti: utility, solidity and beauty.

Utility is rethought as ergonomics, in a world where there will be less and less energy, the question of ergonomics—the radical saving of materials and energy—is fundamental; solidity is interpreted in light of the question of resilience; and finally, beauty becomes the expression of a sober world, where these other principles are perfectly articulated.

F. D. In France there's a debate going on about 5G. Shortly after the demand for a moratorium on the development of this technology presented by representatives of the left and the environmentalist party, the President of the French Republic, Emmanuel Macron, ironically characterised opposition to the innovation of 5G to the "Amish model" and the "oil lamp".

In your opinion, can the technological innovation promised by 5G lead to the solution to the challenges of contemporary ecology?

S. M. Retro-innovation cannot be overlooked as technology that can improve our world and its legibility. That is to say, to examine a whole set of techniques that are cheaper in terms of energy and materials as potentially relevant, thus acquiring great importance for today.

The reference to the Amish is particularly clumsy, as there is certainly much to learn from them.

It's no coincidence that Wendell Berry, a great philosopher of the evolution of agriculture and livelihoods in the United States, dedicated an entire chapter of his book The Unsettling of America to a model that in many technical aspects could represent the Amish community.

The absurdity is to pit these two against each other in this way. In my opinion, making fun of the opponents of 5G by referring to the Amish model reveals the ignorance of their contribution to culture.

F. D. One last question, in your opinion can we oppose a development of society dictated by technological progress and development? Or is technological development essential for survival?

S. M. We are "embarqués", we can only acknowledge that we are forced to flee forward, in a continuous competition on these issues. The question we can ask ourselves is whether it's possible to resist all this, but we can also doubt that this rush will continue for a long time. With all the doubts we can have on this subject, there are obviously two fundamental issues as to energy and materials. Until proven otherwise, all these technologies, notwithstanding their claims of energy efficiency, consume an extraordinary amount of energy in their design, in their operation and in the materials they use, many of which are fossil resources, resources that we want to save. Consequently, even those who are more enthusiastic about technological development can no longer afford to ignore the need to think about what we can call sobriety of energy and materials.

There's no salvation thanks to technology, unless a catastrophe occurs and technology is freed from the need to find these energy and material resources in a reasonable radius around our planet. Otherwise I really don't see how optimistic we can be about technology.

F. D. Thank you very much for your answers. Professor Borsari will conclude this interview.

A. B. Rather than conclude, I prefer to add a few words and ask a final question.

I very much appreciated the approach that Professor Marot showed us: not to oppose technology but to accept the situation in which we are all "embarqués", quoting Pascal's words.

Yes, we are all "embarqués" in this situation, but we must be careful to avoid a shipwreck.

The risk of sinking is very high. What are the means that we can use to avoid this fate?

There are two points that I'd like to emphasise in this regard that intrigued me:

The first point concerns inevitability: we must not be too optimistic and try to save ourselves with technics, but at the same time we must try to use tools that allow us to avoid the most significant damage and at the same time optimise the use of technics.

So once again, use "bottom up" practices, use tools that can allow us to understand, repair and recycle, all indications that to me seem very useful.

In this regard I have a reflection: at the suggestion of Federico I read a few pages of the philosopher Pierre Caye from his book Durer. There I found very precise words on the problem of the relationship between technology and the city. Pierre Caye writes very incisively that technology puts the city at its service, and that this is inevitable. I was struck by how Caye underscores this risk, which more than a risk is already a reality.

The second point I would like to highlight concerns the gap between what the German philosopher and sociologist Simmel had pointed out between subjective culture and objectified culture. The idea that it would be very nice if we could know the technical objects, but is this possible?

Is the knowledge we have sufficient to achieve this goal?

There's a gap that's growing larger every day. How can we halt this trend of our age, which is increasingly marked and probably inevitable?

S. M. I think Pierre Caye's book Durer is a very important book, which has the merit of directly addressing the questions you ask today regarding our mastery of the evolution of technics. What I find very interesting in Caye's book is his systematic way of reinterpreting the fundamental elements of the production system:

The concept of "capital" redefined in light of the notion of heritage, that is subject to protection and control, the notion of "work" reinterpreted in light of repair and maintenance, whether in the field of IT, agriculture or industry, and to consider the importance of maintenance necessary to make development sustainable; and finally "technics", analysed in light of the paradigm of architecture. It's rather powerful as a reflection device: architecture as a paradigm of technics has the objective of expanding time and space. The architect must give time to space and space to time. It's a hard concept to understand unless you think about it until it becomes an almost evident fact. It's a question of giving the development of technics a conscious direction, a rational criterion.

Is it possible to imagine and design a technology that gives time to space and space to time?

It seems essential to me to include Alberti's tradition of architectural reflection at the centre of the technical one, which Caye translated into French with Françoise Choay. Alberti wrote a fundamental reflection on what a city is, or how it works. I believe that Pierre Caye's work is indeed an important one.

With regard to the second question, it's the question of intelligibility, or the legibility of the "milieux" and the worlds in which we live as an important criterion of their duration in the sense in which Pierre Caye uses this term.

I believe that we must introduce legibility, as Kevin Lynch suggests, in the functioning of our worlds, and this is why I often prefer to use the word world rather than the word city, which is always a bit reductive.

Provided however that we use the term world in the plural. Today we use the word world only in the singular, but the manufacture, management and maintenance of worlds in the plural seems to me the condition for a world in the singular to develop peacefully.

The world as a whole is very difficult to read and we aren't aware of the functioning of the worlds we exist in, they have no legibility. We must therefore make every effort to restore a new legibility to our local environment, without forgetting that it participates in a world and in a global technical development.

A. B. Thank you. We could discuss this at length and I understand your answers well, on the other hand the problem with digital is the prospect of having a digital double to make the reality of the worlds that you have evoked comparable or similar to.

This is the risk of the digital dimension overtaking the physical dimension, both in the city and in the world or the worlds in general. This dominance seems to be the destiny of our age, not just a simple trend through which our effort can be achieved and results obtained. But this is the pessimism of reason and intelligence eclipsing the optimism that, as you mentioned, is required for this effort. Thank you again.

Dario Costi Arch. PhD, is Full Professor in Architectural and Urban Design in the Department of Engineering and Architecture at the University of Parma. He is the Director of the interdisciplinary teaching and research project "Smart City 4.0 Sustainable LAB", which is promoted by the University of Parma in cooperation with the Emilia-Romagna regional government and the public digital infrastructure Lepida, and involves the Universities of Ferrara, Bologna, Modena and Reggio-Emila, Piacenza as well as local Authorities and stakeholders. He is member of the PhD Program in Civil Engineering and Architecture at the University of Parma and at the Faculties of Architecture of the University of Palermo and Sapienza University of Rome. He is active both as a professional and as a researcher in the architectural and urban design field. He is the co-owner of the MC2 Architecture Studio in Parma. Since 2007 he is serving as President of Parma Urban Center, of which he is also the co-founder. Involved in numerous national projects for fostering education in architecture and urban design, Professor Costi is the author of many Italian publications, and Editor of two Italian book series published by MUP Editore, Parma. Since 2020, he is the editor of the international book series "The City Project—Strategies for Smart and Wise Sustainable Urban Design", published by Springer. He is also serving as a lecturer for the Program Master Housing at Roma Tre University, in Rome, and for other Master Programs at the University of Parma, Trani, Lecce and Matera. Since 2018 he is serving as Scientific Director of the Higher Education Courses in Urban Strategic Design at the LUM School of Management in Bari. Since 2021, as part of this collaboration, he has been developing the guidelines for the urban regeneration of Italian cities for the SNA—National School of Administration.

Federico Diodato is a PhD in architecture and design cultures from the University of Bologna in co-operation with the Université Paris Est Créteil—Val de Marne (UPEC).

Andrea Fanfoni is a PhD in architectural and urban design from the Department of Engineering and Architecture at the University of Parma.

Emanuele Ortolan is a PhD in architectural and urban design from the Department of Engineering and Architecture at the University of Parma.

Research Finding and Directions Identified by the Smart City 4.0 Sustainable LAB

Giovanni Leoni, Andrea Borsari, Guya Bertelli, Michele Roda, Dario Costi, Gabriele Lelli, Carlo Alberto Nucci, Roberto Menozzi, Gianluigi Ferrari, Sergio Duretti, Francesco Leali, Francesco Pasquale, and Marko Bertogna

Abstract

This chapter contains 11 writings in which the research directions of the cultural souls belonging to the Smart City 4.0 Sustainable LAB are illustrated from very different disciplinary perspectives. These texts provide a starting point for a sustainable development capable of bringing together all disciplines that deal with cities.

1 Smart City: The Decrepitude of the Ideology of Innovation. A Reflection by Giovanni Leoni

See (Fig. 1).

The subject of the Smart City has not been consolidated and most likely will not be consolidated as a field of scientific research. Maybe it was and is a field of action-research.

But for better or for worse the importance of the Smart City derives above all from its being a project. Like all the most effective projects today, able to branch into the various specialist disciplines and capture the attention of politics,

GiovanniLeoni · A. Borsari · C. A. Nucci
Alma Mater, Studiorum University of Bologna, Bologna, Italy
e-mail: giovanni.leoni@unibo.it

A. Borsari
e-mail: a.borsari@unibo.it

C. A. Nucci
e-mail: carloalberto.nucci@unibo.it

G. Bertelli · M. Roda
Politecnico di Milano, Milan, Italy
e-mail: guya.bertelli@polimi.it

M. Roda
e-mail: michele.roda@polimi.it

D. Costi (✉) · R. Menozzi · G. Ferrari
University of Parma, Parma, Italy
e-mail: dario.costi@unipr.it

R. Menozzi
e-mail: roberto.menozzi@unipr.it

G. Ferrari
e-mail: gianluigi.ferrari@unipr.it

G. Lelli
University of Ferrara, Ferrara, Italy
e-mail: lllgrl@unife.it

S. Duretti
Lepida ScpA, Bologna, Italy
e-mail: sergio.duretti@lepida.it

F. Leali · F. Pasquale · M. Bertogna
University of Modena and Reggio Emilia, Modena, Italy
e-mail: francesco.leali@unimore.it

F. Pasquale
e-mail: francesco.pasquale@unimore.it

M. Bertogna
e-mail: marko.bertogna@unimore.it

D. Costi and G. Leoni (eds.), *Smart City: A Critical Assessment*, The City Project 6,
https://doi.org/10.1007/978-3-031-51288-9_4

Fig. 1 Lungo la Via Emilia (Along the Via Emilia) © Photo Alessandra Chemollo

the Smart City is a project of a narrative nature and not a figurative matrix. There is no iconography of the Smart City, only stories.

The structure of the Smart City as a project is very traditional: one speaks of a future city, a better city that does not yet exist but could exist. A topos of the modern project.

As a project, Smart City has: (1) a vision; (2) an articulation in specialisms; (3) an economic structure; (4) a political and communicative dimension.

The vision is elementary and not very innovative if we think of the two centuries that preceded us: technology will improve the quality of life in cities. This task seems to be entrusted to two areas in particular: (1) the management of big data; (2) technologies aimed at reversing the ongoing climate disaster. History shows that the theorem technical innovation = raising the quality of life in cities is not a foregone conclusion, because if we can ascribe quality improvements to technical innovation—or rather to the little control we have exercised over it—we can also ascribe to it many of the difficulties, if not many of the tragedies, that the Smart City project intends to remedy.

Thus, the fragility of the Smart City project derives from its total reliance on specialised technological innovation without cultivating grounds for sharing and comparing different areas. More precisely, from the inability to develop non-specialised tools, conceptual frameworks, that are able to place limits and measures on the specialised actions of technological innovation that, by their legitimate nature, arise without limit or measure aside from the competitive ability to innovate.

The inability to build instruments to control and measure technological innovation then dangerously borders on an unwillingness to set limits and measures if we consider the economic structure of the Smart City project. Because it is clear that, as a powerful vector of the economic dynamics of any action focused on technological innovation, the Smart City project tends to be subordinated to purely financial logics, often getting reduced to a smartwashing narrative.

The only terrain where the Smart City project —with this or any other nom de guerre—seems to be able to acquire meaning and dignity by transforming itself from slogans into concrete actions in the city is perhaps the terrain of politics. One could argue that the subject of the Smart City is no longer missing from most city policy programmes, but its inclusion comes with a choice: (1) politics—understood as the action of political decision-makers—can continue to avoid responsibility by relying on the

Fig. 2 Lungo la Via Emilia (Along the Via Emilia) © Photo Alessandra Chemollo

technological optimism of the Smart City narrative; (2) politics can take on the role of guiding and calibrating actions of technological innovation, giving them an instrumental role.

Choosing the second option has significant implications for vision: (1) it means knowing how to go beyond two centuries of unconditional trust in technological innovation as a guarantee of economic and social improvement; (2) it means finding—also in the field of technological innovation—a mediation between the financial economy and the social economy; (3) it means rethinking the role of specialised knowledge with respect to ethics and public engagement.

But above all it means carrying out actions of empowerment, of deliberate self-disempowerment of the institutional political actor and of specialised knowledge—including the now consolidated specialism of participatory processes led by politics—to build uncertain lands not modelled on pre-established narratives or on an imagined future city (and here it must be said that during the pandemic urban imagination as a guide to the real processes of transformation of cities revealed all its fragility and paucity). It is only by creating terrains of uncertainty, fields of listening, of comparison and self-limitation of specialised actors that the smartness of cities can emerge. Returning political action and the action of technological innovation back to the role of acts of service and government of the city that happens, opening up the field of action to creative cultures, to each actor and his or her specific ability to elaborate city cultures.

The smartness that will emerge from a serious questioning of the invisible and potential dimension inscribed in the body of cities, in its constant and prevailing happening, will not be a single, universal, bright, future technological image but a site-specific smartness, not a model that is imposed on fixed standards and indices, but the calibrated and measured construction of places built as common goods (Fig. 2).

2 A Survey on Temporary Citizenship for a Smartness of Potential City. A Reflection by Andrea Borsari

See (Fig. 3).

The idea of overcoming the uniforming, optimistic and non-specific character of technological innovation by opening up to the uncertainty of the real city as a place to intercept the potential for site-specific transformation, so well set out in the position paper by Giovanni Leoni, finds a possible field of deployment in the

Fig. 3 Lungo la Via Emilia (Along the Via Emilia) © Photo Alessandra Chemollo

research on temporary citizenship that has accompanied the participation of the working group of which we are both part in the broader development of a critical vision of the notion of the smart city. Rather than a general disciplinary call concerning the aesthetics of the city, in this context it may be more useful to point out a specific research project in which different competences fall and which concerns what we have defined as temporary citizenship, that is, the confrontation with a dimension of gap or displacement related to the practices that a series of subjects implement within the urban context and which do not find adequate recognition in the formal definitions of citizenship.

The aim is to reactivate the concept of citizenship in a way that is not just a matter of time, but also a matter of place.

That is to reactivate, in a virtuous way, that dimension of the porous city (porosity is a category used by Walter Benjamin in his time, but recently taken up by Richard Sennett in a re-functionalised key) as a way of intervening in urban contexts to reopen communications and connections where there were none, creating a city that is capable of withstanding and enhancing the overlapping and embedding of different functions and social stratifications, enhancing our historic cities, be they medium, small or large.

This dimension of precariousness is installed precisely where there is a gap between formal and substantial rights, where there is a crisis, as it has been defined, of citizenship based on the erosion of formal status, on the precariousness of access to resources with a tendency to oppose and segment an idea of territorialisation of citizenship, with unequal rights and forms of discrimination.

This is the context within which we are developing a research with a working group within the University of Bologna, involving different competences straddling human, social and city sciences, from planning to architectural design, which puts us on the threshold of reflection concerning the observation of the way in which practices incorporate the not yet explicit directions in which the status of citizenship is stressed—in the sense of subjected to unusual deformation—and integrated.

There are a number of research data, including empirical data, that allow us to restore connections within the city. For example, we are working on the forms of temporary residence starting from the observation of the forms of housing of the students in order to redefine the concept of homing itself, also because we come from an experience that saw us subjected to forced confinement.

More generally, this is a series of researches that attempt to question the new frontier that

Fig. 4 Lungo la Via Emilia (Along the Via Emilia) © Photo Alessandra Chemollo

continually opens up on the slabbing and the margin between practices and formal recognition, focusing on the dimension of precariousness and temporariness, with the tendency to become definitively temporary, that characterises this situation and that allows us to recover a porous image of the city and its potential reality (Fig. 4).

3 Home as Infrastructure: New 'Models' for Living in the Post-covid City. A Reflection by Guya Bertelli and Michele Roda

See (Fig. 5).

It sounds like a paradox to connect the concept of 'home' with infrastructure. But, among the many impacts, the pandemic pushes us to radically rethink the topic of living, with wide-ranging consequences that call into question the architecture's traditional boundaries.

In this line 'home' comes out of its singular dimension, enters a logic of complexity and proposes itself in a systemic vision as the 'infrastructure of the future': the physical, social and economic 'backbone' of a country that, in this crisis, has to find the push for strong innovation.

So, 'home'—even more the public one—redeems its own fragility, becoming the key factor in a process that just new design actions can fill with quality.

The concept of 'home-infrastructure' is proposed as an innovative vision able of re-establishing severed ties in the direction of a strong change in consolidated relationships. A plurality of factors contribute to this dimension:

- in the first place, the necessary rethinking of housing environments, so strongly stressed in lock-down periods, which have largely revealed conditions of backwardness and inability to give effective responses to the inhabitants' expectations;
- therefore the requirement to bring the public house back to the center of thought and politics, a condition that claims a today very important jump in scale, capable of reinterpreting the home itself as an integral part of urban fabrics, in a renewed balance among public and private, among opening and closing, among nature and artifact;
- finally, the standing of a gaze oriented towards the most fragile groups of populations, both from an economic point of view and from a socio-cultural one, as the pandemic has exacerbated a dramatic request for adequate and quality social housing.

Fig. 5 Lungo la Via Emilia (Along the Via Emilia) © Photo Alessandra Chemollo

Starting from this assumption it seems possible—and promising as well—to introduce within the debate the concept of 'house-infrastructure', in a perspective that intends the project of the 'house'—especially the public one—no longer within a submissive and defensive logic, but within an open and projective vision, which looks at it as a generating principle of urban form, integrating within the consolidated habitats, regenerating obsolete areas, building new forms of participation for a society that think (and will think more and more) the domestic space as a relational place: between city and society, between built spaces and urban connective, between nature and flows. As an infrastructure, in fact.

These issues are urgent and necessary because the pandemic has shown the fragility of living and of 'house', attracting the gaze of many critics who have drawn a dramatic and unexpected condition: "*Here we are, in full pandemic from Covid-19—all inside the greatest heterotopia ever known [...]—here we are forced to see the rest of the world enclosed in the space of our own home*" (translation by authors).[1] A condition that has pushed an intense debate in architectural culture, from which—even among very different positions—widely shared aspects have emerged:

- the absolute inadequacy of our housing assets in providing coherent responses to the renewed framework of needs;
- the centrality of 'home' in social dynamics, as a factor able of influencing the level of inequalities;
- the suffering of the real estate market, unable to offer, for some sectors of society, appropriate housing solutions, able to respond effectively to a pulverized and fragmented demand, characterized by multiple situations and by countless types of families—ever smaller, Italian or international, locals or 'nomads'.

Starting from these assumptions, it is appropriate, and necessary as well, to outline a new horizon: moving between decline and development means stressing in-between space, where global questions call for local actions and vice-versa, where inertia and resilience are the key-words. From here it is perhaps important to 'start-again'. From those situations signaling their state of vulnerability and weakness: the degraded suburbs of the 'widespread city', the abandoned spaces of post-industrialism, the interstitial areas,

[1] Pierluigi Nicolin, Architettura in quarantena, Skira, Milano, 2020.

the environments 'forgotten' by the most recent planning. In fact, precisely these 'intermediate territories'—marginal places or fragments of a centrality never been completed—are able to inspire, in their being perpetually in a state of potential 'contention', possible opportunities for redemption from a situation of immobility which is already settled.

In fact, it is in these marginal places that 'house' can take on a new 'morphogenetic' role since it is able to trigger «*a generative process that acts for local, limited, discrete surroundings, but induces very extensive structural changes. They affect the qualitative contents and not just the quantitative data, regardless of the dimensional scales of the intervention, since they are based on the relational links that exist between the different components of the urban framework and transcribe the polysemic nature of the space*».[2] It's a nature capable of introducing not just a relational dimension of the house itself—which in this rediscovered relationship between local and global takes on the new meaning of space 'between'—but also a dynamic vision of a significant node between the permanent condition of 'being within' and the variable condition of 'going outwards', within a continuous dialogue based on mutual correlations and interferences.

In this process, it will be necessary to re-elaborate the distinctive principles of living on the basis of new parameters of coexistence among town and society. Today more than ever, in fact, architecture finds itself the need to recover its social mandate, aimed at understanding, in the unexpected urgency, the boundaries of disciplinary ethics and collective duties. Especially for the public 'house', where the reflection must be even more intense: the dissolution of the public space within the domestic walls, together with the parallel hybridization of the private space, urgently require a rebalance of the internal-external relationship, at the moment in which the 'house' seems to take on an increasingly complex and composite form—'plural' we could say—not very open to a homologating reduction but oriented towards the multiplication of identities.

Rather than responding to a series of goals and indicators, the 'home' of the future will have to be articulated on the basis of a complex reaction to some standards, which could be based on five dialectical pairs today strongly involved in the debate on housing:

- Sustainability/Hospitality;
- Creativity/Multifunctionality;
- Accessibility/Connection;
- Security/Inclusion;
- Adaptability/Resilience.

To conclude with the hope of a fertile outcome of the conceptual categories, we could recall various prefigurations over time aimed at exploring spatial hybridization, overcoming conventional compartmentalizations in favor of a multiple relational, communicative and partici patory connective and therefore with intrinsically infrastructural requirements. From the often utopian origins in the radically innovative intentions of the layouts (so evident in the daring experiments of the Sixties) to the requests oriented to emancipation from housing shortages, to the impulses aimed at redesigning the boundaries between individual and collective spaces as social places, we can see precious testimonies of the multiple explorations of new frontiers of plural living. In this sense, it could also be argued that the golden age of Italian public housing was an extraordinary example of 'infrastructuring' of those places where today we would like to open a season of regeneration, hand in hand with the commitment to abandoned areas, urban voids, obsolete and degraded places.

What seems to gradually distinguish the most recent experiences, however, is a growing awareness of the level of complexity of the inhabited areas (in a physical-social-economic sense), especially in metropolitan areas, which pushes to the research for strategies of high spatial integration, going beyond existing diaphragmatic separations.

The challenge therefore seems very difficult today, since it is aimed at a brief recomposition of

[2] Sergio Crotti, 1991, translated by the authors.

Fig. 6 Lungo la Via Emilia (Along the Via Emilia) © Photo Alessandra Chemollo

the different elements competing in the definition of the concept of 'house-infrastructure', with the hope that the requests, advanced by various parties for more just, green and healthy cities (referring to the recent UN-Habitat dossier) fully involve living, and mainly the 'public house' sector.

But this is still a goal today.

'House-infrastructure' is for now a design vision, a look to the future, an attempt to update—in the face of a changed awareness of living, perhaps still too shaken to assume definitive solutions—principles and paradigms to face the new 'spirit of the time' (Fig. 6).

4 The Seventh Dimension. A Reflection by Dario Costi

See (Fig. 7).

I have entitled this paper *The Seventh Dimension* because, as demonstrated by much of the scientific literature in this field, the smart city is always evaluated and estimated through six dimensions[3], but in these categories of analysis the city of public spaces, the city of architecture, the city of the physical configuration where the community lives is almost entirely lacking.

Joseph Rykwert told us clearly that in antiquity the idea of the city was represented by monuments and foundation rites that bound people to the city through its built symbols.[4] Today, what remains of this rite of belonging? What does the idea of the city mean today? It means having physical places that we all recognise as references of our urban identity, of our emotional relational stability.

The built city and its architecture are necessary for the city of life.

Stefano Zamagni recalled how the Romans spoke of the perfect combination of *civitas* and *urbs*, that is, between the city of souls and the city of stones, as if they were representations of a dual dimension, that of people and that of space. They are two sides of the same coin, two realities that are continuously intermingled, verified and overlaid. If we imagine returning to the concept of the city of the Romans, we must imagine a co-presence of a physical city and a city of life, of

[3] Francesco Manfredi, *Smart Community. Comunità sostenibili e resilienti (Smart Communities. Sustainable and resilient communities)*, Cacucci Editore, Bari 2015.

[4] Joseph Rykwert, *L'idea di città, Antropologia della forma urbana nel mondo antico (The Idea of the City. Anthropology of the Urban Form in the Ancient World)*, 1963, recently republished in Italy by Adelphi Milan, 2002.

Fig. 7 Lungo la Via Emilia (Along the Via Emilia) © Photo Alessandra Chemollo

souls, of people. In reality, if we are to address this fundamental theme seriously, we must recognise that in city evaluation systems, which imply a recognition of objective values, this co-presence is reduced to almost nothing.

In a recent study that we carried out,[5] we selected 20 ranking systems for evaluating Smart Cities, from the world scale to the regional scale of Emilia-Romagna, where 183 indexes were identified, derived from 975 indicators that we catalogued and traced back to the 6 dimensions (smart governance, smart living, smart people, smart mobility, smart economy and smart environment). Of the 975 indicators, only 19 (just 2%) concern architectural, urban design aspects and relate to the quality of public spaces, denoting a major difference with the ancients, who envisaged a symbiosis, a continuous dialogue, estimated at a ratio of 50% *civitas* and 50% *urbs*. This difference points to a cultural, even more than a substantive, problem that arises from an approach that disregards civilisation as a finite place that has contributed to the process described by Bauman,[6] causing communities to liquefy.

The mission is therefore to reconstitute the community around the new possibilities of interconnection capable of triggering those processes of connective democracy mentioned by Derrick De Kerckhove, as well as the processes of re-appropriation of physical space for a new *Right to the City*.[7]

The post-war city is a city that has become progressively peripheralised. It has disintegrated, it has alienated people, it has not enabled them to be together. The city of the future needs to be completely reorganised, reconstructed and reconnected, echoing the thought of Giovanni Maria Flick, who attributes to the city the role of "social formation" envisaged by the Constitution, as an instrument of personal growth, the bridge between the individual and the institutions. The *Right to the City* thus becomes a constitutional

[5] Francesco Manfredi, Dario Costi, *Community Regeneration Masterplan. The Five Dimensions of Sustainability: Guidelines For European Cities*, Springer, Berlin, 2023.

[6] Zygmunt Bauman, *Modernità liquida (Liquid Modernity)*, Laterza, Bari 2002. An updated perspective is captured in the short interview *Futuro liquido, Società, uomo, politica e filosofia (Liquid Future, Society, Man, Politics and Philosophy)* edited by E. Palese, Edizioni AlboVersorio, Milan 2014.

[7] Henri Lefebvre, *Il diritto alla città (The Right to the City)*, Marsilio editori, Padua, 1970.

right to be claimed in order to demand a better city, a city that creates the conditions for truly integrated social living.

Urban design themes are now more critical than ever. When we think of the city of Zero Land Consumption, Renaturation and Urban Regeneration we have to envisage a city where public spaces can be reconnected through urban parks, intermediate spaces, fragile landscapes, places that can be reclaimed to reconnect the entire great legacy of parks and squares that already exist. Digital technology has certainly made relationships between people more virtual, but the physical city has also greatly increased this dispersal.

I have been discussing the themes of the *City of People 4.0* for some time, drawing on the idea of the City of Man that was connected to Adriano Olivetti's concept of Community—the concept of overcoming the struggle between the material and spiritual spheres—while also trying to understand how the fourth industrial revolution can be above all a service to the community.[8]

In coining the term *fourth industrial revolution*,[9] Klaus Schwab reminds us that the more we think about the goals we can assign to these tools, the more powerful will be the application of technology and the debate on the *culture of technology*, which has perhaps always been somewhat lacking in the West but is now more essential than ever.

Everything is indeed changing and we must prevent the affirmation of that view of technology described by Emanuele Severino.[10] The prospective scenarios point to a very interesting future.

Academics working on autonomous driving remind us that in a few years' time, perhaps 80% of parking will no longer be necessary, 80% of accidents can be avoided, and pollution will also be reduced by 80%. What I call the 80% LESS rule opens up an important prospect of reclaiming public spaces in the city. We all know that today's urban voids are largely car parks, spaces designed for a city of cars. Hopefully, these areas can be reclaimed by civilised living and handed back to the people.

Therefore, today we must begin to reflect on how these technologies can not only improve people's lives but also incentivise, stimulate and bring about an urban regeneration of public spaces as places for a community that can and must be reconnected. So let us imagine a Wise City, the *City of People 4.0*, as a place to consider the reconnection of public spaces, the design of community spaces on the surface level and the creation of the digital infrastructure system at the underground level, in order to harmonise the technologies and architecture of the city, but also to recover the relationship between the city of souls and the city of stone (Fig. 8).

5 Smart Small City. A Reflection by Gabriele Lelli

See (Fig. 9).

The digital revolution is an opportunity to rethink our cities, cities which talk differently with other urban elements and where real and digital could merge into one single identity. Future urban transformations have to address digital and technological innovations. In urban design, this means rebalancing both urban resources and urban quality of life objectives. Within this framework, the new innovative tools become new design resources, while the redefinition of urban quality remains a fundamental step to realize a contemporary idea of the city with its urban values.

Our work with the Next City Lab group deals with it, conducting applied researches on forthcoming urban issues. Doing applied research, landing ideas and innovations in the physical territory while verifying the results, is the most suitable approach to face this period of sweeping changes and take advantage of the intertwining

[8] Adriano Olivetti, *Città dell'uomo (City of Man)*, first edition 1960, republished by Edizioni di Comunità, Ivrea, 2015.

[9] Klaus Schwab, *La quarta rivoluzione industriale (The Fourth Industrial Revolution)*, Franco Angeli, Milan, 2016, p. 16.

[10] Emanuele Severino, *Il destino della tecnica (The Destiny of Technique)*, Rizzoli, Milano, 2009.

Fig. 8 Lungo la Via Emilia (Along the Via Emilia) © Photo Alessandra Chemollo

Fig. 9 Lungo la Via Emilia (Along the Via Emilia) © Photo Alessandra Chemollo

bonds between real city and digital tools. Just by way of example, using Digital Twin solutions could be interesting if they create applications that improve physical urban space and quality of life. As the Wien Digital Twin platform, realized and used to manage public space and offer a new way to involve their citizens.

There are three main areas to leverage urban transformation processes. Firstly, the electronic devices that have so much contributed to change our lives and our cities. Smartphones, sensors, drones, electric scooters, and self-driving cars, the list is constantly updating.

The second area concerns the behaviors of city users and communities, which generally change urban life without affecting city form. This is an outstanding lever for change, but it must start from the bottom, from local

Fig. 10 Lungo la Via Emilia (Along the Via Emilia) © Photo Alessandra Chemollo

communities, to have the potential to deliver its positive impacts. Only in this way can be created effective smart communities, where blockchain technologies are deployed to generate new economic and social value. The growth of other needs and common goals will slowly contribute to the creation of other urban behaviors.

Finally, the most visible lever for change of our cities, i.e. the infrastructures and the urban form. In this field, the transformations are often designed as minimal interventions on urban fabric to increase urban quality and trigger new synergies, as we can figure out from the many reuse and urban regeneration projects. They tackle public space, roads, bike paths, new urban services, and a new relationship with the natural environment. All urban dimensions are intertwined with each other to make new synergies. RAISE>UP, a spin-off from the University of Ferrara, aims to find these synergies.

According to this approach, our applied research is focusing outside the largest cities of the Emilia Romagna region, in the sprawled lands of small and mid-sized towns. In these areas, the "building mass" is different from cities making it necessary to explore other strategies to turn them into Smart Towns. Within this process, critical thinking is a key point to design effective solutions with a positive impact on local communities and the environment. Furthermore, we have to consider the territory of small and mid-sized cities as a network of places where their local identity can be valued, preserved, and improved.

In conclusion, urban transformations for small towns, that often are characterized by good living standards, will have as main objective the citizens' quality of life while maintaining their bell tower and local identity. This challenge can be pursued by innovating urban services and the meeting opportunities, which are currently lacking, taking advantage of the new enabling solutions of the current digital revolution (Fig. 10).

6 The City's Potential for Eenergy Transition. A Reflection by Carlo Alberto Nucci

See (Fig. 11).

The EU Mission on Climate Neutral and Smart City combines the achievement of climate neutrality with the implementation of the Smart City model. A city today cannot be 'only' smart, it also requires climate neutrality, i.e., the achievement of net zero greenhouse gas emissions. To achieve this double objective, Europe envisages the development of various financial

Fig. 11 Lungo la Via Emilia (Along the Via Emilia) © Photo Alessandra Chemollo

instruments and regulatory interventions, accomplished by using the indications gathered through a bottom-up approach, along with the introduction of a Climate City Contract, ideally signed by national or regional authorities and possibly by the European Commission. A multilevel governance is indeed considered an indispensable element for the success of the Mission. The progressive electrification of consumption in various sectors, favoured by the growing diffusion of generation powered by renewable sources, and the push towards the creation of energy communities envisaged by the RED II and IEM European directives, underline the growing importance that the electricity system will play in this context. A massive use of information and communications technologies (ICT) for power transmission and distribution is necessary to cope with the uncertainty of renewable sources and with the increased complexity of the control and operation of the electricity system. In recent years, an increasing number of small generation units have been connected to the part of the electricity system—the distribution networks, the one close to the end users who are now also able to produce significant quantities of electricity—where only consumption was originally foreseen. The term consumer is therefore often replaced today by the self-explanatory term prosumer. Thanks to the diffusion of ICT, electricity, thermal, gas, and even water supply networks are expected to be managed in a more coordinated way to exploit their synergies: one of the most interesting technical challenges for the Smart City is hence such a smart sectors integration. What briefly outlined above, offers more than one element to justify what has been already argued for some years, namely that the Smart Grid is the first enabler for the implementation of the Smart City concept, especially with reference to energy sustainability, allowing the efficiency improvement of traditional networks and services for benefit of citizens. The enhanced ICT infrastructure, the increasing development of renewable generation connected to distribution networks, the progressive electrification of some sectors, the availability of smart devices by consumers and prosumers, and the rapid spread of Internet of things are the relevant characteristics of the Smart City.

It is now generally accepted that the climate emergency must be addressed in cities: they cover about 3% of the planet's soil but produce over 70% of all global greenhouse gas emissions. Cities are growing rapidly; in Europe, it is estimated that by 2050 almost 85% of Europeans

Fig. 12 Lungo la Via Emilia (Along the Via Emilia) © Photo Alessandra Chemollo

will live in cities. Cities are also the "laboratory" in which the decarbonisation strategies of energy, transport, buildings, and even industry and agriculture coexist and intersect. As the density of infrastructures and their use is greater in cities than in the countryside, there is also greater potential for intersectoral integration of complex infrastructures to achieve a more general smart grid paradigm. A convincing example of what has just been stated is represented by Energy Communities, formed by citizens' associations, commercial activities or companies for the production and sharing of electricity from renewable sources. They own indeed a significant potential for the accomplishment of the energy transition in our country and represent the core of smart districts, capable of transforming themselves into a so-called Positive Energy Districts (Fig. 12).

7 Into the Future with Open Eyes. A Reflection by Roberto Menozzi

See (Fig. 13).

The topics discussed in the interviews transcribed here offer a wealth of indications and suggestions that may help us reflect critically on the *smart* and—hopefully but not automatically—*wise*[11] future of our cities. In particular, the field of *autonomous drive* and *intelligent mobility* can serve as a paradigm for the several technological transformations encompassed by the *smart city* transition. The first common trait of these ongoing transformations is the synergic *convergence* of a number of technologies—some of them very new, some not so much—on the activities characterizing our lives of social and productive beings (e.g., mobility), aimed at an increase of their efficiency, safety, and sustainability. The key word here is *convergence*: while the relevant technologies are not necessarily novel or revolutionary, what is novel and revolutionary is that they recently came to cooperate in the testbed of our cities raising a sort of *perfect storm* where each of them becomes an enabling and accelerating factor for the others.

The penetration of this variety of new tools aimed at making our activities simpler and more efficient, however, does not come without a price to pay, nor is it risk-free: we should be aware of

[11] "For wisdom is the property of the dead—A something incompatible with life", W. B. Yeats, *Blood and the Moon*. That is to say, being *smart* is easier for us humans than being *wise*.

Fig. 13 Lungo la Via Emilia (Along the Via Emilia) © Photo Alessandra Chemollo

this price and of these risks lest the *smart* city should become in the end a *smarting* city. In this regard, one of the points most deserving of attention is our progressive and often unwitting waiving of shares of our autonomy, freedom, privacy, and responsibility.

Taking the case of autonomous drive and intelligent mobility as an example—but it would not be hard to find other cases-in-point—a scenario in which vehicle-to-vehicle (V2V) and vehicle-to-infrastructure (V2I) communications allow continuous and pervasive tracking of individual movements should raise—I believe—some Orwellian concern; nor is of little consequence our decision of relinquishing to hardware and software tools sophisticated as they may be—ethically grounded decisions such as those a driver happens to make, often instinctively, when facing unexpected and dangerous contingencies. This sort of concerns is compounded by another one: technologies that are conceived, developed and deployed with a goal—among others—of increased sustainability, such as those concurring to foster more rational and efficient mobility, end up posing themselves new sustainability challenges. For example, we have to consider the problem of upscaling the capacity of the communication and data management infrastructure to accommodate the gigantic amount of V2V and V2I data, as well as the energy demand and environmental impact of the distributed mega-computers and data centers that will control and regulate the large mobility infrastructures.

For these reasons, we should appreciate the value of every opportunity that draws our attention to this kind of cost-benefit analysis.

This vigilant and critical mindset should obviously be far from any kind of obstructionist or luddite temptation, which—on top of being plainly wrong—would be perfectly inane. More than twenty-five centuries ago someone with a keen eye wrote about us: "and now nothing will be restrained from them, which they have imagined to do",[12] thus ringing a very fore-sighted alarm bell about the dangers of our technological *hybris*.

We nevertheless owe it to ourselves and to the generations to come to cultivate this sense of awareness and vigilance to guide us toward well-balanced and humane choices. The progress of technology is ripe with dangers if it is not paralleled by the progress of our ability to critically evaluate its overall impact on our society, wherein we must strive to be *citizens*, not merely

[12] Gen. 11, 6 (KJV).

Fig. 14 Lungo la Via Emilia (Along the Via Emilia) © Photo Alessandra Chemollo

consumers or cogwheels in the productive machine—indispensable as these two roles are.

At the end of the day, our vow can perhaps be summarized paraphrasing the last words of the emperor Hadrian in Marguerite Yourcenar's novel: *Let us try, if we can, to enter into the future with open eyes* (Fig. 14).

8 ICT as the Enabler of a Virtuous Circle in a Smart City. A Reflection by Gianluigi Ferrari

See (Fig. 15).

The cities of the future are expected to be secure, safe and reliable. Smart cities can be defined as "innovative cities that use ICT (Information and Communication Technologies) and other means to improve quality of life, efficiency of urban operation and services, and competitiveness, while ensuring that they meet the needs of present and future generations with respect to economic, social, environmental, as well as cultural aspects".[13]

Digital transformation is crucial to identify the directions that a smart city should follow to shift from being an isolated environment to an intelligent municipality.[14,15] Let us recall that digital transformation is the final outcome of the digitalization process. In the context of smart cities, the steps towards digital transformation are the following:

- Digitization: process of conversion or coding of information in digital form, so that the same information can be managed by computers. More generally, integration of Information Technology (IT) in everyday tasks.
- Digitalization: use of IT or digital technologies to introduce changes inside the existing processes. IT is the enabler to improve the main activities within the entire value chain. In the context of a smart city, the value chain

[13] Simon E. Bibri, John Krogstie, *Smart sustainable cities of the future: An extensive interdisciplinary literature review*, Sustain. Cities Soc. (2017) 31, 183–212.

[14] Mervi Hämäläinen, *A Framework for a Smart City Design: Digital Transformation in the Helsinki Smart City*, in Entrepreneurship and the Community: A Multidisciplinary Perspective on Creativity, Social Challenges, and Business; Springer International Publishing: Cham, Switzerland, 2020; pp. 63–86.

[15] Katarina Tomicic Pupek, Igor Pihir, Martina Tomicic Furjan, *Smart City Initiatives in the Context of Digital Transformation—Scope, Services and Technologies*, Management, 2019, 24, 39–54.

Fig. 15 Lungo la Via Emilia (Along the Via Emilia) © Photo Alessandra Chemollo

encompasses multiple layers: from single citizens to the entire urban community.
- Digital transformation: implementation of the technological transformation, i.e., the most pervasive phase of the evolution that a smart city has to undertake. It identifies a new way of rethinking of the city, a deep change necessary to exploit fully the opportunities which come from the multiple available technologies.

To make the cities of the future intelligent, the first step is represented by the definition and deployment of a reliable connectivity infrastructure, supported by heterogeneous networking, allowing information exchange in the most flexible way.[16] A smart city should improve the life quality of citizens, e.g., through the use of (Internet of Things (IoT)-oriented (or, more generally, System of Systems (SoS)-oriented) technologies,[17] with the final goal of maximizing the efficiency of the offered services. This should also include efforts to make the best use of existing resources (e.g., managing mobility, congestion, pollution, and urban food production) to maximize the safety and security of citizens.[18,19]Such digital transformation is expected to provide European cities with the ability to interact and support each other efficiently, thus fostering European autonomy at the urban level and paving the way to more secure and safer European urban scenarios. Furthermore, a European effort in developing the technologies needed by a smart city will reduce the dependency on other (non-European) providers. RUGGEDISED is an H2020 EU project where six European cities are joining forces to accelerate the path towards a sustainable future by creating model urban areas[20]: three cities act as "lighthouse cities"

[16] Laura Belli, Antonio Cilfone, Luca Davoli, Gianluigi Ferrari, Paolo Adorni, Francesco Di Nocera, Alessandro Dall'Olio, Cristina Pellegrini, Marco Mordacci, and Enzo Bertolotti, *IoT-enabled smart sustainable cities: challenges and approaches*, Smart Cities, vol. 3 (2020), no. 3, pp. 1039–1071. https://doi.org/10.3390/smartcities3030052.

[17] Matthias Lederer, Juliane Knapp, Peter Schott, *The digital future has many names—How business process management drives the digital transformation*, in Proceedings of the 2017 6th International Conference on Industrial Technology and Management (ICITM), Cambridge, UK, 7–10 March 2017; pp. 22–26.

[18] Emanuele Crisostomi, Robert Shorten, Fabian Wirth, *Smart Cities: A Golden Age for Control Theory?*, IEEE Technol. Soc. Mag. (2016), 35, 23–24.

[19] Sam Musa, *Smart Cities—A Road Map for Development* IEEE Potentials (2018), 37, 19–23.

[20] H2020 EU Project RUGGEDISED, https://ruggedised.eu/home/.

Fig. 16 Lungo la Via Emilia (Along the Via Emilia) © Photo Alessandra Chemollo

and three cities act as "fellow cities" (one of the fellow cities is Parma).

The Strategic Research and Innovation Agenda (SRIA) 2022[21] provides a perspective on Electronic Components and Systems (ECS) considering the following layers.

- Foundational technologies: the concept of SoS plays a key role in the development of complex systems, such as a smart city. However, other aspects (such as components, module, system integration and embedded software) are relevant.
- Cross-sectional technologies: most of the aspects of this layer are relevant: (i) artificial intelligence, edge computing and advanced control; (ii) connectivity, as we already mentioned above; (iii) quality, reliability, safety and cybersecurity.
- ECS Key application areas: most of the considered areas apply to a smart city: (i) mobility (vehicle traffic management is always a city's priority); (ii) energy (an efficient smart grid allows the urban authority to have an accurate real-time overview of the energetic situation and, possibly, an accurate forecast); (iii) health and well-being (citizens need to remain active as much as possible and, if fragile, need to be supported); (iv) agri-food (urban gardens are relying on smart agriculture ideas and applications) and natural resources; (v) digital society.

Digital society should really be enabled by a smart city, and vice versa. Therefore, we can identify ICT as the enabler of a virtuous circle:

- a smart city supports a digital society, providing to citizens new and effective tools to improve their life quality;
- a digital society allows a smart city to introduce new technologies, which can be effectively exploited by citizens (Fig. 16).

9 The Role of Digital Infrastructures. A Reflection by Sergio Duretti

See (Fig. 17).

What is the role and the right impact of digital infrastructure and services in a full development of the potential of a smart city and more in general the whole territory? How can a pervasive

[21] Strategic Research and Innovation Agenda (SRIA) 2022, available at https://efecs.eu/publication/download/ecs-sria-2022.pdf.

Fig. 17 Lungo la Via Emilia (Along the Via Emilia) © Photo Alessandra Chemollo

and widely distributed infrastructure strengthen or decrease the road for a more sustainable and wiser city and improve a better quality of public, private and nonprofit services?

The indicators detailed in ISO 37120—Sustainable cities and communities, indicators for city services and quality of life—are the international reference point for sustainable cities with 80 indicators in 19 categories. They can help us to evaluate the general relevance of digital infrastructure to measure smartness and sustainability of a territory.

The analysis shows how 36 indicators in 15 categories are related to digital topics confirming a wide and multidisciplinary coverage. This is true for the economy as well as for energy, for mobility as well as for governance.

The category 18 devoted to Telecommunication indicators points 3 indicators:

- percentage of the city population with access to sufficiently fast broadband as measurement of digital divide of citizens to access to (smart) working, to (distance) education and learning, to (advanced) content platforms;
- percentage of city area under a white zone/dead spot/not covered by telecommunication connectivity as measurement of digital divide of territory;
- percentage of the city area covered by municipally provided Internet connectivity as measurement of public initiatives against digital divide.

Starting from this last indicator in the region Emilia-Romagna, in the first years of the new century, adopted a strategy for a strong public initiative. Using European structural funds in coherence with the Lisbon Strategy launched in 2000, since 2002 was designed and made an optic fiber network for connecting all the Municipality of the region. This network was called Lepida, in honour of ancient Roman consul Marco Emilio Lepido, the builder of Via Emilia. Some numbers give an idea: in December of 2022 the Lepida Network has reached 134,000 Km of fiber optic laid with 5400 optic fiber access points for connecting public buildings (municipal buildings, libraries, theatres, museum, hospital, public health centers and laboratories, universities, research center, schools). If the network described reaches the goal of the third indicator, another infrastructure is closely related to the second indicator: a South Radio Backbone is capable of reaching even the most remote areas with adequate performances; it qualifies as a further asset useful to provide solutions against the Digital Divide and management of emergencies with 226 radio transmission sites.

Fig. 18 Lungo la Via Emilia (Along the Via Emilia) © Photo Alessandra Chemollo

Regarding the first indicator, since 2017 the project EmiliaRomagnaWifi covers areas on the regional territory, with the provision of a free service for citizens in public buildings and places. It provides Ultra Wideband WiFi connection homogeneously and without registration with 10,800 access points, 1 for 408 inhabitants.

Finally 2 others initiatives is running:

- an IoT Network for Public Administration based on LORAWan protocol, is available to both PAs and private citizens, associations, companies, aiming to allow the collection of data from thousands new sensors, placed at their own chosen positions;
- an infrastructure for Industrial Areas, now 90 in the whole region, as provided by a Regional Law, with ultra broadband for companies.

All these assets described are managed by Lepida ScpA, the ICT in-house public company, owned by 442 local public bodies (Region, Municipalities, Universities, Health local public company and other public structures).

It is important underline as final consideration that this "platform" (not only technological bur also organizational, relational, social—in a word cultural) is not a sufficient condition for doing a smart city and territory but is the best way for enabling multidisciplinary ideas and proposal and make them achievable (Fig. 18).

10 Autonomous and Connected Mobility within the Smart City Framework. A Reflection by Francesco Leali and Francesco Pasquale

See (Fig. 19).

It is fair to say that the more autonomous the vehicles will be from the driver, the more dependent they will become on the city and its digital infrastructure.

The interconnection between CCAM (Cooperative, Connected and Automated Mobility) and infrastructures will grow proportionally with the introduction of driver assistance systems, that are gradually filling the market. Based on a scale that conventionally goes from 0 to 5, where 0 stands for no driving assistance, while 5 represents a completely autonomous driving, at the state of the art it is allowed to circulate on open roads with level, but prototype are testing already at level 5 in confined ODD (Operative Design Domain). The next stage 3 already involves the

Fig. 19 Lungo la Via Emilia (Along the Via Emilia) © Photo Alessandra Chemollo

transition from V2V (vehicle to vehicle) to V2X (vehicle to everything) technologies, thus making the vehicle interact with the surrounding environment.

While in the nineteenth and twentieth centuries the infrastructures of modernity heavily impacted the environment with concrete and steel for the construction of roads, bridges, tunnels and stations, nowadays the data infrastructure for computer science and information technology are largely intangible like the wave frequencies in which they travel. Yet it no longer make sense to split infrastructures in the two categories of tangible and intangible: all the orthodox infrastructures now imply some form of digital equipment and every digital technology will always rely on a physical network and interface devices for its usage.

Overcoming this dichotomy leads to consider an integrated design of triple interactions among human, environmental and technological networks, each of them being digital and physical at the same time. Human networks are all the interpersonal relationships, starting from the family and thus the dwelling, including affections, friendships, work relationships, associations, institutions, etc... The environmental networks are those that determine the eco-systemic balance of the land, such as green and blue infrastructure, ecological corridors, biospheres and different site-specific naturalistic compound. At last, but first as carbon footprint impact, the technological infrastructure are those typical of the anthropic occupation of land. They primarily concern the management, as production and/or distribution, of energy, water, transport, and waste.

The mobility pattern should better integrate with the other network systems not only from a technological, but also an organic perspective on the city, in terms of a global complexity that exceeds the sum of the single disciplines. Recognizing the primacy of people over machines, *civitas* over *urbs*, the progressive and yet inevitable adoption of technologies based on artificial intelligence will always be subordinated to a scale of socio-cultural values of the community to which they will be proposed.

In order to match these values, the information technology underlying autonomous driving should enable three major achievements. The first one undoubtedly concerns inclusiveness, as the process of extending the right to mobility as large as possible to fragile or disabled people. The second one complies with sustainability, as traffic reduction is crucial in terms of both reducing emissions and regulating the discipline of the use of the soil resource. The third goal

Fig. 20 Lungo la Via Emilia (Along the Via Emilia) © Photo Alessandra Chemollo

faces the essential matter of security, in the double meaning of the reliability of an ethical algorithm that would preserves the physical safety of driver, passengers and other road users, and in that of data cybersecurity, a very broad discipline that ranges from privacy to the prevention of hackering.

In the field of civil transport, connected and driverless vehicles available on demand, without not only the costs of the ownership and the maintenance—as already happens today with some financial loans—but also free of the inconvenience of the parking and a garage, could be the key factor for the turnover of mobility models from car-ownership-centred to MaaS (Mobility as a Service), offering a widespread and tailor made transport service for a wider range of people or goods. Innovation will be played out on ongoing processes, crucial for the necessary background knowledge, as well as on new markets for innovative products and services, which will allow to break down further categories such as private/public, individual/collective transport.

New vectors will change the infrastructure and thus the environment in which they will move, both at the scale of the micro-mobility of the last mile and in the so-called *15 minutes city* models, and in the longer journeys on urban and regional routes, which must be guaranteed and improved precisely in order not to fall back in the epigone of those urban models that generated healthy neighbourhoods on one side and ghettos on the other (Fig. 20).

11 Experiences of Integration Between Driverless Vehicles and Urban Infrastructure: The Case Study of the MASA Open Lab—Modena Automotive Smart Area. A Reflection by Marko Bertogna

See (Fig. 21).

MASA's mission is to prototype the technologies for the mobility of the future, which are largely linked to information technology, artificial intelligence, computer science and computer vision, vehicle/man, vehicle/machine and vehicle/infrastructure communication technologies, as incremental process of enabling equipment not only for mobility, but for all the smart city realms. The open lab operates in the real

Fig. 21 Lungo la Via Emilia (Along the Via Emilia) © Photo Alessandra Chemollo

context of a city area, testing level 5 autonomous driving applications on different types of vehicles, ranging from racing cars to small drones or delivery bots. In fact, while the technological avantgarde is still challenged in the racing compartment, the technological transfers will affect multiple vectors for urban mobility.

In terms of urban infrastructure, along the test routes of the area have been installed multiple intelligent cameras, with integrated chips on which algorithms have been written, in order to detect obstacles and other road users. Due to a deep respect towards privacy issues of images and all the provisions related to the GDPR legislation, the technology chosen is of the *on the edge* type, i.e. without remote communication as the data is encrypted by the camera itself before being sent. The possibility of real applications is linked to the time in which it is possible to carry out the transmission and processing the data, thus constituting a key factor in guaranteeing a safe and smooth running: to date MASA is the only research centre that can certify an operational time less than 100 ms.

Although the current applications of autonomous driving are of an exclusively experimental step, the technological goal for commercial use is not far as it might look, given the continuous progress in the field of artificial intelligence and neural networks that govern data, and consequently by that the vehicles. Obviously, however thin the margin of error still exists, its resolution remains the constraint for the homologation and the mass production of driving assistance systems of a higher level than 2.

Assuming that it is therefore only a matter of time, the adoption of these technologies will have a disruptive effect on the urban landscape, comparable to that of the car itself, which has shaped most modern cities. Architects and landscapers, who have the task to imagine urban structures for the next decades, will take the challenge to integrate the innovations of mobility in the larger frame of the city (Fig. 22).

Fig. 22 Lungo la Via Emilia (Along the Via Emilia) © Photo Alessandra Chemollo

Giovanni Leoni, Giovanni Leoni (1958) is Full Professor in History of Architecture at the University of Bologna where he directed the Department of Architecture, the Doctorate in Architecture and Design Cultures and was a member of the Board of Directors. His research focuses on the theory and practice of design between the nineteenth and twentieth centuries with particular attention to models of creativity, the relationship between Personality and Anonymity and the role of architectural design in social and political processes. He edits the journals Histories of Postwar Architecture and European Journal of Creative Practices in Cities and Landscapes (with Carole Hein, TU Delft). He coordinated, for the University of Bologna, the H2020 project ROCK Regeneration and Optimisation of Cultural heritage in creative and Knowledge cities and is part of the Bologna unit of the PNRR project CHANGES Cultural Heritage Active Innovation for Sustainable Society, Spoke 7, theme: The historic Italian city as a model for sustainable urban development. He has published extensively on the themes of architecture and the city in Emilia-Romagna in the second half of the 20th century.

Andrea Borsari, Full chair in Aesthetics at the University of Bologna—Department of Architecture, where is also faculty member of the Ph.D. program in Architecture and Design cultures. His research focuses on aesthetic experience, cultural memory and its pathologies, urban culture and philosophical reflection on city, architecture, design, landscape and cultural heritage. Among his recent publications: Georg Simmel, Hans Blumenberg, and philosophical anthropology, in The Routledge International Handbook of Simmel Studies, London—New York, Routledge, 2020; Body, in Glossary of Morphology, Zürich, Springer International Publishing, 2020; Georg Simmel et le Champ Architectural. Sociabilité urbaine, paysage et esthétisation du monde, Milan—Paris, Editions Mimésis, 2020.

Guya Bertelli, is Architect and Full Professor at Politecnico di Milano, where she teaches Architectural and Urban Design. President of the Course of Studies in Architecture (2013–14) and in Sustainable Architecture and Landscape Design (2015–2019), she collaborates with Italian and international schools. She holds lectures and conferences, participating in numerous educational initiatives. Since 2010 she has been directing the OC OPEN CITY International Summer School at POLIMI Piacenza Campus. Within DASTU Department, she's working and researching in the "Department of Excellence" multi-year program on territorial fragility. Member of DOPAUI, ASP (Alta Scuola Politecnica, 2012–2015) and ASA (Advanced School of Architecture, 2013–2014). In parallel with the teaching commitment, she conducts theoretical and design research. Since 2019 to 2022 she was expert member of the "Struttura Tecnica di Missione" of the Italian Minister of Infrastructures and Transport, working as a delegate for urban regeneration in fragile contemporary territories and as a member of the High Commission of the PNRR Innovative Program for the Quality of Living (PINQUA).

Michele Roda, Architect (1978) carries out teaching and research activities at the AUIC School of Politecnico di Milano, dealing mainly with architectural and urban design topics. Since 2010 he has been coordinator of OC OPEN CITY International Summer School at POLIMI Piacenza Campus, where he organizes many international workshops and events, dealing with transformations of contemporary environments and landscapes. He publishes books and essays and he writes in architectural newspapers and magazines, mainly as member of the editorial board of "Il Giornale dell'Architettura". As a professional, he is author of numerous projects of social housing, a line of activity he's strongly working about also in research.

Dario Costi, Arch. Ph.D., is Full Professor in Architectural and Urban Design in the Department of Engineering and Architecture at the University of Parma. He is the Director of the interdisciplinary teaching and research project "Smart City 4.0 Sustainable LAB", which is promoted by the University of Parma in cooperation with the Emilia-Romagna regional government and the public digital infrastructure Lepida, and involves the Universities of Ferrara, Bologna, Modena and Reggio-Emila, Piacenza as well as local Authorities and stakeholders. He is member of the Ph.D. Program in Civil Engineering and Architecture at the University of Parma and at the Faculties of Architecture of the University of Palermo and Sapienza University of Rome. He is active both as a professional and as a researcher in the architectural and urban design field. He is the co-owner of the MC2 Architecture Studio in Parma. Since 2007 he is serving as President of Parma Urban Center, of which he is also the co-founder. Involved in numerous national projects for fostering education in architecture and urban design, Professor Costi is the author of many Italian publications, and Editor of two Italian book series published by MUP Editore, Parma. Since 2020, he is the editor of the international book series "The City Project—Strategies for Smart and Wise Sustainable Urban Design", published by Springer. He is also serving as a lecturer for the Program Master Housing at Roma Tre University, in Rome, and for other Master Programs at the University of Parma, Trani, Lecce and Matera. Since 2018 he is serving as Scientific Director of the Higher Education Courses in Urban Strategic Design at the LUM School of Management in Bari. Since 2021, as part of this collaboration, he has been developing the guidelines for the urban regeneration of Italian cities for the SNA—National School of Administration.

Gabriele Lelli, Architect and Associate Professor of Architectural and Urban Design at Ferrara University, Faculty of Architecture. Urban innovation is part of his research interests. He has taught at Venice University (IUAV di Venezia), at the Mendrisio Academy of Architecture, in Switzerland and in Spain. Graduated with Adolfo Natalini, he has collaborated with Mario Zaffagnini, Massimo Carmassi, and Peter Zumthor. He founded the design firm Cristofani and Lelli Architetti and LBLA + partners. In 2014 set up Next City Lab research group, active on Smart and Ethic City issues. He is Director of the Master Programme Innovation Manager, Real City and Digital City.

Carlo Alberto Nucci, Graduated with honors in electrical engineering from the University of Bologna, Bologna, Italy, in 1982. He is a Full Professor and Head of the Power Systems Laboratory of the Department of Electrical, Electronic and Information Engineering "Guglielmo Marconi", University of Bologna. He is an author or coauthor of over 370 scientific papers published in peer-reviewed journals or in proceedings of international conferences. Prof. Nucci is a Fellow of the IEEE and of the International Council on Large Electric Systems (CIGRE), of which he is also an Honorary member, and has received some best paper/technical international awards, including the CIGRE Technical Committee Award and the ICLP Golde Award. From January 2006 to September 2012, he served as Chairman of the CIGRE Study Committee C4 System Technical Performance. He has served as IEEE PES Region 8 Rep in 2009 and 2010. He has served as Editor-in-Chief of the Electric Power Systems Research journal (Elsevier) from 2010 to 2022. He has served as the President of the Italian Group of the University Professors of Electrical Power Systems (GUSEE) from 2012 to 2015. He is presently serving as Italian Representative in the Horizon Europe Mission "Climate-neutral and Smart cities", and as a member of the Technical Scientific Committee of the Regional Energy Plan of Emilia Romagna Region, Italy. Prof. Nucci is Doctor Honoris Causa of the University Politehnica of Bucharest and a member of the Academy of Science of the Institute of Bologna. He is also serving as chair of the International Conference on Lightning Protection, ICLP.

Roberto Menozzi received the Laurea degree (cum laude) in Electronic Engineering from the University of Bologna in 1987, and the Ph.D. degree in Information Technology from the University of Parma in 1994. He is Full Professor of Electronics with the Department of Engineering and Architecture, University of Parma. His recent research activities are mostly focused on the numerical simulation of thin-film solar cells, and smart energy systems. He has published about 190 papers in refereed international journals and international conference proceedings. He is a Senior member of the IEEE.

Gianluigi Ferrari (http://www.tlc.unipr.it/ferrari) is a Full Professor of Telecommunications at the University of Parma, Italy, where he coordinates the Internet of Things (IoT) Lab (http://iotlab.unipr.it/) in the Department of Engineering and Architecture. Since 2019, he is a member of the Scientific Council of INSIDE Industry Association and of the board of the Scientific Council of Consorzio Nazionale Interuniversitario delle Telecomunicazioni (CNIT). He is an IEEE Senior Member. Since 2016, he is co-founder and President of things2i Ltd.(https://www.things2i.com/), a spin-off company of the University of Parma dedicated to IoT and smart systems. things2i is currently developing city2i®, a platform for smart city data integration and analysis.

Sergio Duretti, Born in 1962 in Turin. After scientific studies, he began to work in the private sector (bank and media companies). Since 1996 he worked for Public Administration as Chief of Telematic City Project of Municipality of Modena (1996–99), as Chief Innovation Manager (1999–2007) and CEO (2008–2016) of CSP, Piedmont Supercomputing Center. Since 2017 he worked in Lepida ScpA in Bologna as Director of Digital Transformation (2017–18), Digital Welfare (2019–20) and actually Director of Strategic and Special Actions. He took part in 26 European Projects. He has been enrolled in national register of journalists since 1984.

Francesco Leali is Full Professor of Design and Methods of Industrial Engineering at the University of Modena and Reggio Emilia, holds a Ph.D. and a Master's Degree in Mechanical Engineering. He is president of the Master's Degree in Advanced Automotive Engineering, Coordinator of the Interuniversity Committee of MUNER and coordinator of the strategic research line "Automotive Academy UNIMORE". He is scientific coordinator of IDEA research lab, investigating Computer Aided Design for Automotive, Model Base Design—(GPS and GD&T), automation and industrial robotics. He is the author of international scientific publications in scientific journals and international conferences, for which he received some awards.

Francesco Pasquale is an Architect, Adjunct Professor of Landscape Architecture at the University of Ferrara, Department of Architecture, and research fellow at the University of Modena and Reggio Emilia by the open lab MASA—Modena Automotive Smart Area. As professional he worked as project leader in the dutch offices MVRDV and Mecanoo before to establish his own firm Brenso, which has been awarded for projects in the field of landscape and urbanism. He is author of several international scientific publications as well as he has been speaker in national congresses on landscape, mobility and environment. He has been founder and president of the association GArBo—Giovani Architetti di Bologna and he is now counsellor of the Professional Association of Architects, Landscapers, Planners and Restoration of the Province of Bologna, spokesperson for environment and infrastructure.

Marko Bertogna is Full Professor and leader of the HiPeRT Lab of the University of Modena and Reggio Emilia. His main research interests are in High-Performance Real-Time systems, especially based on multi- and many-core devices, Autonomous Driving and Industrial Automation systems. In 2008, he received a Ph.D. in Computer Sciences from the Scuola Superiore Sant'Anna of Pisa. He authored more than 100 papers, receiving the 2009 Best Paper Award for the IEEE Transactions on Industrial Informatics, and 9 other Best Paper Awards in first level international conferences. He coordinated multiple EU and industrial projects, securing more than 10 MEuro in funding for his research group. He is Senior Member of the IEEE, and Stakeholder Member of HiPEAC. He is CEO and founder of the academic spinoff HiPeRT Srl.